BIOENERGY RESOURCES
Planning, Production and Utilization

BIOENERGY RESOURCES
Planning, Production and Utilization

Pradeep Chaturvedi
Secretary
Indian Association for the
Advancement of Science
New Delhi

Foreword by
Dr. M. S. Swaminathan

CONCEPT PUBLISHING COMPANY, NEW DELHI - 110059

ISBN 81-7022-549-3

First Published 1995

Published and Printed by
Ashok Kumar Mittal
Concept Publishing Company
A/15-16, Commercial Block, Mohan Garden
New Delhi - 110059 (India)

FOREWORD

Energy is a key input in every field of human life and endeavour. Rich countries consume both more energy and more natural resources. For example, the United States is the biggest single user of natural resources. In 1989, the U.S. consumed 4.5 billion metric tonnes of natural resources—or 18 metric tonnes per person. With less than one-third the population of India, the U.S. consumes nearly three times as much iron ore and 12 times as much petroleum. U.S. energy consumption is so large that its per capita carbon-dioxide emissions are 19 times that of India.

As a result of the high energy consumption, the industrialised countries also contribute more to green house gas emissions. The United States and the former Soviet Union together contribute 32.8 per cent of the total atmospheric impact of current global greenhouse gas emissions, 19.1 and 13.6 per cent respectively. The European Community nations would rank third at 12.4 per cent. The top six countries contribute 55.8 per cent of global emissions.

The book "Bioenergy Resources - Planning, Production and Utilization" written by Mr. Pradeep Chaturvedi is therefore a timely contribution. The book provides comprehensive information on the progress made during the last decade in mobilising bio-energy sources. Mr. Chaturvedi has also dealt with the various constraints in the use of bio-energy. This book is not only an excellent source of analysis of recent developments but also a valuable input in public policy formulation. We are indebted to Mr. Chaturvedi for this labour of love.

Madras.

DR. M.S. SWAMINATHAN
President
M.S. Swaminathan Research Foundation

CONTENTS

ACKNOWLEDGEMENTS

Inspiration, help and cooperation from a number of experts and colleagues, were crucial in the compilation of this book.

My grateful thanks are due to Dr. M.S. Swaminathan for sparing time to write the Foreword.

I gratefully thank the Bioenergy Society of India, for approving publication of this book in the memory of Dr. O.P. Vimal, its Founder Secretary. I am also grateful to the family of Dr. O.P. Vimal for the permission to dedicate this book in his memory. Help from his son, Mr. R.K. Vimal, in this regard was important. I also gratefully acknowledge support of Shri Ashok Kumar Mittal, Managing Director, Concept Publishing Company for undertaking publication of this book.

I would like to thank Dr. Gustavo Best, Senior Energy Coordinator FAO Rome for his help and inspiration.

There may be quite a few other persons who have not found their mention in the acknowledgement. But this dosn't mean I am not grateful to them. I am.

New Delhi
November 17, 1994

PRADEEP CHATURVEDI

INTRODUCTION

Renewable energy sources were promoted in the country in the early eighties. Relevance of bio-energy resources was considered significant and, therefore, it was decided to lay the foundations of a forum where scientists, technologists, entrepreneurs, administrators, policy-makers and bio-enthusiasts, interested in the growth and development of bio-energy resources in the country, could meet, discuss and analyse various issues. This was the genesis of the Bioenergy Society of India (BESI), which was born out of the loud thinking of the undersigned along with late Dr. O.P. Vimal and Dr. Dilawar Singh. As Dr. O. P. Vimal was then actively involved, both professionally and officially, with the development of the bio-energy resources in the country, it was considered appropriate to hand over the responsibility of running the Society to him. We can now proudly look back on the dedicated efforts of my esteemed friend, Dr. Vimal, who was instrumental in laying a sound foundation for the Bioenergy Society of India.

BESI is an organ to project bio-energy resources in the right perspective. Its relevance to Indian conditions is crucial as it draws attention of the policy-makers to provide support to the bio-energy programme. BESI, therefore, focussed its attention more on application and on issues that required coordinated effort. Also BESI tried to involve a large number of application oriented (action research experts) rather than only the basic science research personnel in the country. The perspective was very clear. The dissemination of technological developments had to aim, in the initial phase, at improving the energy crop productivity; and at the end-use equipment efficiency in the rural and remote areas. The linkage between agriculture and wasteland development was also realized.

I have specifically mentioned the Bioenergy Society of India as most of the papers included in the present volume were presented at different times at the BESI Conventions or related conferences.

As a Founder-Member of BESI, and now as its Vice-President, I consider it my duty to project the policy issues for the development of bioenergy resources in India in a cost-effective optimized fashion. During this journey, since the foundation of BESI, I have had occasions to make presentations at national and international seminars, conferences and conventions. My effort has always been to relate policy issues to international developments. Therefore, these presentations were mainly projections of technical, economic and social issues and not of scientific advancements alone. All these papers have been collected in this volume to reflect a wide scenario of the growth pattern in the bioenergy sector during the past 10 years. A Chapter on "Bioenergy Options for India" has been specially added to project the policy interventions necessary to achieve higher cost effective utilization of bioenergy resources, initially by the year 2000 and in the long-term by the year 2020. It is my firm conviction that appropriate policy measures can ensure atleast 3 mtoe fuel oil replacement and 2,000 mw of installed power generation capacity based on bio-fuel systems. The revolution can be in terms of a higher penetration of bio-energy resources in the industrial and urban sectors. The contribution in the traditional rural domestic and rural industry sector shall continue but more efficiently. All this will require an appropriate government policy on the one hand and the enterprise, managerial skills and business acumen of the industry on the other. Future developments will not only be through the government channels, but these cannot even take place without government support and intervention. A number of actions have no precedence and the industry would like the government to share the risk till soundness of investment is proven. Growth and development of this sector will be in the national interest and, therefore, government should consider providing all help and support.

"Bioenergy Resources: Planning, Production and Utilization" is the outcome of the journey started with my friend, Late Dr. O. P. Vimal, who breathed his last serving the bioenergy sector. I consider it a privilege to dedicate this book to his memory, with the permission of his family.

November 17, 1994 PRADEEP CHATURVEDI

1

BIOENERGY OPTIONS FOR INDIA

Environmental concerns after the UNCED have given a new thrust to bioenergy resources and its impact is also felt in the country. Energy cropping is not a traditional practice but has been quite akin to the agricultural cropping except for the fact that the produce is not given a true tradeable value. By and large the understanding in India is that biomass resources either in terms of fuelwood or else agro-residue and cowdung are available free of cost to the poor people in rural areas. The fuelwood has also been considered a natural growth in forests.

The concept of "afforestation" and "energy plantation" has been varied and this has been realized in the country, more so, after the global environmental concerns at the UNCED in 1992. Various independent studies in India indicate the average plant survival rate in the forestry sector to be 11 per cent ; in the agro-forestry sector up to a maximum of 50 per cent and the average plant survival rate in the energy plantation under the schemes of the Ministry of Non-Conventional Energy Sector to be 69.69 per cent. Such results have been possible only due to proper selection of plant species; appropriate inputs of water and fertilizers; and by proper management practices. All this is cost intensive and, therefore, the "energy plantation" is costlier than 'afforestation'; but all higher costs are compensated by higher returns from better produce.

Policies to encourage development of the bio-fuels industry have to be closely coordinated with national agricultural programmes and with efforts to restore degraded lands. Unfortunately, the wasteland reclamation itself is in the pilot phase and is being operated in selected districts in the country. Large demonstrations in different agro-climatic zones only can build confidence in the users. As a result of the UNCED, the Global Environment Facility (GEF) of the World Bank, the United Nations Development Programme (UNDP) and the United Nations Environment Programme (UNEP) have started allocating large funds for initiating major projects and programmes based on biomass in the developing countries. The major countries where such projects can be executed successfully in the South are Brazil and India. This funding has to be effective in India also.

Bio-fuel Resource Potential and Conversion Technologies

The present potential in India is as follows:

Bio-fuel Resource Potential

Bioenergy Plantation Potential

J. Davidson, in his paper on Bioenergy Tree Plantations in the Tropics-Ecological Implications and Impacts, published by IUCN in 1987, says, "It appears reasonable to state that prospects for commercial success are better if bioenergy plantations are established only with those species on those sites which will yield not less than one per cent and preferably over two per cent solar ECE (corresponding to about 10-20 t/ha/yr above-ground biomass production respectively) over an rotation of five years. (ECE is energy conversion efficiency). (For a detailed reading please see Annexure).

J. Davidson also recommends, "For high productivity, commercial bioenergy plantations normally should be established in areas having an annual rainfall of not less than 1,000 mm and preferably greater than 1,250 mm.

Water use by eucalyptus plantations has been a controversial subject particularly in India. Available scientific evidence indicates eucalyptus have no more effect on the water table than several other species of commonly planted trees and, on the basis of unit weight of dry biomass produced, eucalyptus are relatively efficient in their use of water and they maintain biomass production under conditions of moisture stress."

Opinions differ, but energy plantations in general cannot be categorized to harm the ecological balance.

Suitability of bioenergy plantation as sustainable biomass production can be aimed at the wasteland available in the country which account for 93.69 million ha of the non-forest land and 35.89 million ha of the forest land. The position is that against 140 million ha of land under cultivation and fit for cultivation, we have got nearly 130 million ha not available for any good productive purpose.

Moreover, the degraded non-forest land is in the following composition:

(i)	Saline and Alkaline	17.17 mha
(ii)	Wind Eroded	12.92 mha
(iii)	Water Eroded	73.60 mha

Of these degraded lands, only 7.17 mha of the saline and alkaline lands can be reclaimed and put under energy plantation at a reasonable cost which will again be higher than the present estimates for the normal soil. A good deal of energy plantation will require a support of Rs. 40,000 to 60,000 per ha for a period of five years. Consequently, firewood productivity of up to 15 tonnes per hectare per year, on a production cycle of five years, can be attained.

Although experiments at the Biomass Research Centre of the National Botanical Research Institute, Lucknow, and the Biomass Research Centres at Bhubneswar and Madurai have achieved fuelwood productivity levels of up to 36 tonnes per hectare per year, this cannot be generalized. However, depending on the inputs and the financial assistance made available for management of the plantation, a continued sustained productivity of up to 15 tonnes per hectare per year can be expected.

In case all the 7.17 mha of saline and alkaline land is put under energy plantation at a rate of 15 tonnes per ha per year, it will be feasible to produce about 110 million tonnes of firewood on degraded lands.

There is also a possibility of successful energy plantation on wind eroded lands but their reclamation costs will be much higher. Such plantations may cost anywhere between Rs. 50,000 and Rs.75,000 per ha for a period of five years.

If these two categories of degraded lands are taken together, then an effort can be made on 20.09 mha of waste land that can produce about 300 million tonnes of firewood every year. Such a possibility of regeneration of the degraded land will take a time frame of 15 to 25 years and, therefore, such targets have to be fixed for the year 2010 to 2020. However, if an effort is made now, then in the first biological cycle an effort can be made to grow energy plantation over 1-2 mha of land by the year 2000. Accepting the quantum jumps of up to three times in the growth pattern, as observed in the cereal production under the Green Revolution, the target of 20 mha can be achieved by the year 2020.

Quality Planting Material

Such an approach besides requiring a financial assistance will also require quality planting (genetic) material. The Ministry of Non-Conventional Energy Sources intends to set up Biomass Research Centres in each one of the 15 agro-climatic zones that will be involved in the quality genetic material production. This first-line genetic material is parallel to the foundation seeds in agriculture and need to be multiplied through nurseries which have to serve as breeders. Thus, a linkage between the research centre and the ultimate user has to be established through an appropriate mechanism of quality planting material producers.

Involvement of private industries to grow energy plantation on such degraded lands is a suitable and appropriate option in the country. They will be able to run these plantations on commercial lines avoiding the government's financial commitment.

But there are certain administrative lacunae, for example, most of the degraded land is under the possession of the government or panchayat. Under the present Land Ceiling Act, nobody can hold large land holdings and even if the government or the panchayat agrees, an economically viable size of land holdings for energy plantation cannot be leased out.

Agro-Residue Potential

In a MNES sponsored survey, conducted by the National Productivity Council in 1987, it was revealed that about 321 mt of crop residues and 50 mt of agro-industrial bi-products were available in the country. Another MNES sponsored survey in 1988 estimated that about 94 mt of agro-residues and industrial wastes are not used at all for any purpose. S.K. Singh, et al, in their Paper on 'Role of Agro Residues in Meeting Energy Demands in India' presented at the IAAS Workshop on 'Biodiversity and Bio Energy' held on Oct 6, 1993 have shown that 94 mt of agro-industrial residues and wastes have shown that 94 mt of agro-industrial residues and wastes have a potential to replace about 25 mt of furnace oil or else a total of 90 mt of coal assuming 3.5 kgm of briquettes can replace 1 kgm of furnace oil.

Conversion Technologies

Briquetting

Briquetting is a commercially viable proposition. Substitution of furnace oil or coal in industrial and commercial establishments is a lucrative and possible option. Furnace oil consumption in boilers is 10 mt per year. It is feasible to produce 3.5 mt of briquettes every year, to save 1 mt of fuel oil, by the year 2000.

Such an effort does not call for any extra infrastructural cost for biomass production; as the raw material is already "wasted agro-residue". Moreover, all the investment for conversion into briquettes shall be borne by the industrialists.

Gasification

Substitution of furnace oil through the gasification route is also feasible.

Biomass gasification is already an attractive proposition for decentralized power generation up to 100 kW range. Systems of even higher capacity have been developed. Application in the thermal mode is found interesting for some of the industrial applications especially for curing of tobacco and tea leaves. Technical feasibility aside, it is an attractive option that competes costwise with the conventional power systems. This route also opens other opportunities for use of woody biomass in the form of both agro-residues and fuelwood. Specific species have been developed by the National Botanical Research Institute for burning in the gasifier.

This application for the thermal mode use in the industry and for the electric mode use at the rural level provide ideal solutions under certain specific conditions. In the case of the industries that have the capacity to grow their own energy plantations the gasifier route needs to be pursued vigorously.

D.K. Khare et al have projected the possibility of replacement of one mt of furnace oil by gasifying 3.5 mt of fuelwood per annum, burnt at a gasification efficiency of 75 per cent. If all this firewood is to come from a linked energy plantation, then an area of 0.175 mha would be required for sustainable growth. D. K. Khare et al have assumed a yield of 20 tonnes per ha per year after five years of plantation. A feasible programme of one tonne of furnace oil replacement per year can be planned for the year 2000.

Biogas Applications

Cowdung has normally been taken as a poor man's fuel and that has been the reason why biogas applications for income generation were not promoted in the initial phase of the National Biogas Development Programme. The technology has been already perfected at a low scale of operation, say, up to 4 cubic metre size plant. Even large-size plants have been developed up to 85 cubic metre capacity but for household cooking only. Commercial application of biogas should be paid due consideration. Most appropriate application would be in the thermal mode utilization of biogas for heating. Dairies where cowdung concentration exists may be promoted to use biogas for heating to produce ghee or promote such activities as the bakery unit where the oven may be run on biogas. In view of the uncertainty about the biogas production and fluctuation in summer and winter, it would be most appropriate to develop applications in the hybrid mode. That is to say, the oven in the bakeries should be capable of being run on firewood even when biogas is not available.

The potential of biogas generation can be fully tapped only with its commercial exploitation through private sector undertakings. As this application is still to be developed, it is difficult to project specific target. But an effort at saving one mt of oil may be relevant by the year 2000.

Urban Waste

Urban waste and industrial waste pose a serious environmental problem due to lack of proper disposal facilities. However, two possible routes identified are through incineration and dry fermentation. Technologies do exist for such applications but not much work has been done in India. The bitter experience of the urban waste plant failure in Delhi has almost put a stop to development of urban waste treatment in the country. There is a strong possibility of developing up to a 1,000 mw through the urban and industrial waste by the year 2000.

Co-generation

Co-generation is an important route with a vast potential for generating surplus power to the tune of 4,000 mw through biomass-based co-generation systems in the sugar industry alone. The MNES Task Force on Cogeneration has observed that more than 95 per cent of the sugar mills produce their own electric power (22 ATA, 330°C) from bagasse-fired boilers in back-pressure turbo generators, the exhaust of which is utilized as-process system. Realizing the economic advantages of surplus power generation, a few sugar mills have initiated steps to install high-pressure boilers (45 ATA) with the definite aim to generate surplus power either to feed their own ancillary units such as mini paper plants, distilleries, chemical units, etc., or for sale to grids.

The Ministry of Non-Conventional Energy Sources has floated a project scheme that will support establishment of co-generation for production of 4 mw or more surplus power per plant through a new system instalment. For another category of projects which wish to utilize existing facilities with minor modifications for the generation of 1 to 4 mw of surplus power, commercial loans are suggested. Based on this, by the end of 1997, installation of an additional power generation capacity of 300 mw is recommended.

This is one of the most promising options and one cannot rule out the possibility of reaching a target of 1,000 mw of additional installed capacity through the co-generation route by the year 2000.

Improved High Efficiency Cook Stoves

The early effort for creating an awareness about the utility of high efficiency cook stoves has already disseminated this option on a large scale. This programme should really now be the common man's programme where continuous efforts on demonstration may be promoted.

New Areas of Biomass-to-Energy Conversion Technologies

Carpentieri et al (1992) and Goldemberg et al (1992) while discussing the bioenergy conversion technologies have identified five fundamental forms of biomass energy use:

(i) The "traditional" domestic use in the developing countries for household cooking, lighting and space heating. In this role, the efficiency of conversion of the biomass to useful energy has been raised from 5 per cent to 20 per cent.

(ii) The "traditional industrial" use of biomass for the processing of tobacco, tea, pig iron, bricks and tiles, etc., where the biomass feedstock is often regarded as a free energy source. There is generally little incentive to use the biomass efficiently and so the conversion of the feedstock to useful energy commonly occurs at an efficiency of 15 to 30 per cent.

(iii) The "modern industrial" activities relate to thermal conversion technologies. Expected conversion efficiencies are between 30 and 55 per cent.

(iv) Newer "chemical conversion" technologies (fuel cell) are capable of bypassing the entropy - dictated Carnot cycle limit which describes the maximum theoretical conversion efficiencies of thermal units.

(v) "Biological conversion" techniques, including anaerobic digestion of biogas production and fermentation of alcohol.

In general, biomass energy conversion technologies have to deal with a feedstock which can be highly variable in mass and energy density. Size, moisture content and intermittent supply. Therefore, modern industrial technologies are often hybrid fossil fuel/biomass technologies which was the fossil fuel for drying, pre-heating and maintaining fuel supply when the biomass supply is interrupted.

Direct combustion processes have received greater attention in the industrialized world. Technologies are being developed for feedstock use of wood chips, sawdust, bark bagasse, straw, municipal solid waste and wastes from the food industry.

New types of direct combustion furnaces can be divided into two broad categories and are used for producing either direct heat or steam. Dutch ovens, spreader-stoker and fuel cell furnaces employ two stages. The first stage is for drying and possible partial gasification, and the second for complete combustion. More advanced versions of these systems use rotating or vibrating grates to facilitate ash removal, with some requiring water cooling.

The second group include suspension and fluidized bed furnaces which are generally used with fine particle biomass feedstocks and liquids. In suspension furnaces the particles are burnt whilst being kept in suspension by the injection of turbulent preheated air which may already have the biomass particles mixed in it. The fluidized bed combustors, a boiling bed of pre-heated sand (at temperatures of 500° to 900°C) provides the combustion medium, into which the biomass fuel is either dropped (if it is dense enough to sink into the boiling sand) or injected if particulate or fluid. These systems obviate the need for grates, but require methods of preheating the air or sand, and may require water-cooled injection systems for less bulky biomass feedstocks and liquids.

Advanced technology development initiatives have been taken up by scientists in the industrialized countries. Similar work is also necessary to be undertaken in India if our effort is to promote bioenergy use at a fast pace. High temperature gasification based on bagasse is an important application.

Actions for Intensive Use of Bioenergy Resources in India

A few actions at the national level can be conducive to a speedy growth of the bioenergy systems in the country to achieve the target by 2000. These can be summed up as follows :

(i) Immediate term perspective up to year 2000, and long-term perspective up to 2020, should be developed for use of bioenergy resources.

(ii) Bioenergy plantation through private enterprises linked to its utilization needs to be promoted. Such an approach will ensure raw material supply to the main industry and also reduce the infrastructure cost.

(iii) Necessary legislative action must be taken to permit large government land lease for bioenergy plantation, outside the purview of the Land Ceiling Act.

(iv) Fiscal incentives should be made available to the private industries on the same level as for the government agencies.

(v) A package of incentives including low rate of soft term loans should be made available for production and conversion technologies and equipment.

(vi) Intensive R&D should be supported with a view to commercializing the technology under different marketable site conditions.

Whereas in the short-term emphasis should be placed on agro-residue-based technologies, in the long-term energy plantation linked newer conversion technologies should be vigorously pursued.

(vii) A package of technologies for extension should be developed to support quantum jumps of factors of two to three in one biological cycle, as was observed in the case of wheat during the Green Revolution.

(viii) An "all out" effort should be made on different routes including briquetting and gasification for agro-residues; incineration and dry fermentation of urban waste; and co-generation of bagasse in the sugar industry.

Bioenergy in India in 21st Century

Availability of new and efficient combustion and conversion technologies will make bioenergy use economically competitive to fossil energy use on a lifecycle basis. Global competition and opening up of India's economy will certainly force the Indian industry to economize and convert even the industrial and urban waste and effluents into a useful energy resource. Bioenergy resources will certainly play a significant role in agriculture under the rural domestic sector in addition to the emerging industrial sector.

Whereas co-generation in sugar industry will be in the forefront of additional power generation, use of agro waste for briquettes and use of urban waste through dry fermentation for gas production will be close on their heels. This will be supported by many small captive gasifier-based power units in a decentralized and dispersed fashion.

The future impact of bioenergy utilization in the country would lead to improvement in income generation capability in the rural sector as bioenergy production in itself will be labour oriented and, therefore, income generator in the rural areas; and availability of energy will lead to initiation of a number of income generating activities. The rate of penetration of bioenergy in the industrial and urban sectors will be much faster than in the rural sector. The major reason for such transformation would be availability of apppropriate technology for conversion and environmental concerns that will receive attention in urban centres rather than the rural sector.

The realization of the role of energy plantation on carbon sequestration is also building up a lobby for energy plantation. Economic development of degraded lands through energy plantation will be an attractive option and, therefore, future policies for the renewable energy sector, agriculture and wasteland development will certainly have a much more coordinated effort.

An all-out effort at utilization of all sources of bioenergy also include the animal energy, animal waste and urban waste. The government has really to think seriously about the option between letting the available resource go waste and utilizing it effectively. The most effective route would be disposal through utilization as a source of energy. The economics of urban waste utilization would be slighdy tricky. The infrastructure costs will have to be borne by the government and the conversion and utilization should be left for the entrepreneurs to work out the economics.

If the past experience is taken as the basis for future projection then the success stories in energy plantation, and biogas plant utilization have to be seriously taken. Whereas in a survey conducted by the author in 1982-83 in Bulandshahr district near Delhi, the biogass plant utilization was observed to be 25 per cent only, it has gone up to 85 per cent and above in many cases. This was as an effort of awareness creation, training in maintenance and operation, and improvement in technology. There are incidents of 100 per cent trouble-free operation of biogass plants in district Durg in Madhya Pradesh and district Ghaziabad in U.P. for over 10 years.

Similarly, the energy plantation demonstration carried out at the Mass Awareness Convention at the Anjora Farm in district Durg, in 1986, has now resulted in thousands of hectares of energy plantation having been undertaken by individuals on a successful commercial basis.

Under the present circumstances there is little doubt that by the 2000 India could achieve a target of 1,000 mw of installed additional power generation through co-generation in the sugar industry; 1 million tonnes of oil replacement through briquetting of agro-residues; 1 million tonnes of oil replacement through gasification and 1 million hectares of energy plantation.

The successful performance in the formative years of the bioenergy programme in India leads us to believe that bioenergy should be the mainstay for majority of the population as India marches into the 21st century.

2

BIOENERGY PRODUCTION AND UTILIZATION IN INDIA

Biomass growth in India like in many other developing countries has been considered for food production alone. Any suggestion regarding use of landmass for the production of bioenergy plantation always raises the issue of competing needs of food production. Such a discussion goes unabated on all platforms of sustainable development. The large population in India still does not have access to fossil fuels except kerosene for lighting. On the other hand, the urban centres, through inefficient use of fossil fuels, are causing a grave danger to the environment through the Green House Effect from emissions from automobiles, thermal power plants, and industrial boilers, furnaces and ovens. Though the general discussion centres round developing a scenario with ecologically driven development the emphasis is always placed on the use of fossil fuels for speedy development. Development of biomass as an important renewable source of energy gets a back seat.

On the other hand projections for the woody biomass availability and the rate of deforestation due to the use of fuelwood in the rural domestic sector have caused alarm.

Present Situation

Role of Bioenergy in Meeting Energy Demand in India

Fuelwood, agro-residue, urban waste and cowdung provide almost

---

"The Expert Consultation on Biofuels for Sustainable Development: Their Potential as Substitute to Fossil Fuels and in CO_2 Emissions Reduction". Organised by the Food and Agriculture Organization (FAO), Rome, September 13-15, 1993.

50 per cent of India's total energy demand. Figures for fuelwood demand and production, and availability of agro-residues and wastes are as follows:

i. *Fuelwood*
   a. Demand : 166.85 million tonnes per year.
   b. Production : 28.40 million tonnes per year.
   c. Deficit : 138.40 million tonnes per year.

ii. *Total Availability of Agro-Industrial Waste* : 246.00 million tonnes per year.

iii. *Surplus Agro-residue* : 94.20 million tonnes per year.
   Refer Table 2.1 for details of agro-residue.

TABLE 2.1: Surplus Agro-residues in India in 1991-92

Rice Husk	21.46	mt/year
Maize Stalk	3.27	mt/year
Jowar Stalk	8.00	mt/year
Coir Pith and Dust	12.76	mt/year
Groundnut Shell	2.51	mt/year
Bagasse and Sugarcane Trash	18.00	mt/year
Cotton Stalk	24.10	mt/year
Saw Dust	3.40	mt/year
Jute Waste	0.70	mt/year
Total	94.20	mt/year

There are different fundamental forms of bioenergy use but attention in India is always focussed on bioenergy use in the traditional sector for household cooking and heating applications. Firewood has also been used traditional applications in rural industries and for certain industrial applications in urban areas. Traditional industrial applications include tobacco, tea, brick and pottery processing.

Earlier the biomass feedstock was available at a very low cost and, therefore, there was hardly any consideration for efficiency of use, and no incentive was available for its conversion in an efficient fashion. But some steps have been taken to modernize or improve efficiencies of use, for example, by using gas from the biomass gasifier for processing in the traditional tobacco and tea industries. At the same time certain modern industrial applications have been considered in India, for example, biomass briquettes for the industry. In India even a 25 per cent improvement in the use efficiency of firewood can mean a reduction of almost 40 million tonnes of firewood consumption. It is easily said in statistical terms but is very difficult to implement — the problem of efficient utilization of fuels and emission control is related to management practices.

Poor Use in Industry and Household due to Lack of Appropriate High Efficiency Technology

By use of biomass based fuels in the traditional form, the industrialists tend to get a feeling that they are dealing with an outdated energy resource and they would not be able to create an impact of use of modern technology or modern process. Moreover, the concept of economy of scale due to use of concentrated forms of fossil fuel energy also has an adverse impact on bioenergy resources. A switchover from low energy efficiency systems to high efficiency biomass conversion technologies is not simple. The developed countries do not have as high a stake in use of biomass as a source of energy as does India.

The major problems with regard to biomass conversion technologies relate to:

(i) erratic availability,
(ii) variable heat content, and
(iii) variable forms of the feedstock. Supply of the feedstock would be erratic and highly seasonal thereby requiring storage which in itself is a costly proposition. Moreover, the biomass would be of variable energy densities depending on their mass, size, moisture content, etc. (see Table 2.2). Therefore, in every case, site specific technology needs to be developed.

Another aspect is the importance of different sources in the local context. Though biogas holds promise of use in the promotion of rural industry, cowdung is seen as a poor man's fuel and, therefore, use of biogas for any purpose other than cooking is almost prohibited.

TABLE 2.2 : Heat Potential of Various Agricultural Wastes.

Agricultural Wastes	Moisture % (wt basis)	Ash %	Calorific Value Kcal/kg
Rice straw	10.6	20.1	3000
Wheat straw	9.2	18.8	3800
Cotton sticks	12.0	13.5	3300
Maize stalk	11.5	14.2	4700
Maize cobs	8.6	13.8	3500
Bajra stalks	11.2	17.5	3850
Pearl millet husk	-	-	3626
Groundnut straw	12.1	1.3	4200
Castor stalks	-	-	4767
Mustard straw	-	-	3665
Mustard stalk	-	-	4034
Gram straw	9.2	13.2	3950
Moth straw	-	-	3850
Masoor straw	10.1	12.	3810
Moong straw	10.3	12.6	3820
Spent coffee grounds	-	-	5000

A major consideration at the moment in India is to shift the attention of policy planners from the role of woody biomass and biogas only in the "rural domestic energy applications" to "rural industrial and economy building activities". Realization of the role of bioenergy in the promotion of decentralized industries can really mean higher utilization of bioenergy resources.

Role of Bioenergy for Generating Employment through Agro-Industries

The emerging agro-processing sector for a speedy growth in rural area requires a sustainable supply of energy wherein biomass has an important role. Local or rural industry development is also seen in India as a possibility of substantially improving the year round employment which will take away the adverse impact of the seasonality of employment based on agriculture. Under the Poverty Alleviation Programme, the government's effort was mainly to supplement the income of agriculturists and labour to provide year round income generation.

Agricultural production and agro-processing can be complementary to one another and the scenario of local resource use for sustainable development will provide an opportunity for year round employment in villages. That will reduce pressure on transport and will also require low investment for creation of permanent employment. Biomass growing, conversion and utilization will all create employment.

A linkage between bioenergy and rural employment is important and can create up to 100,000 jobs in rural India in the next five years.

Higher Production and Efficient Utilization of Biomass for Energy

Higher Production of Biomass for Energy

Different concepts so far introduced in India for improving the growth of biomass for energy have been the following:

Forest Management

Under the Afforestation Programme, emphasis has always been on timer growing. Planned development started in the year 1951 and substantial improvement in forest preservation started during the Seventh Five Year Plan (during 1985-1990). Progress under afforestation during 1951-1992 has added 18.835 million hectare of area under vegetation. This programme is totally under the government forestry department.

Social Forestry

Under the Social Forestry Programme, the main objective is to produce fuelwood, fodder, small timber and other products to meet the needs of rural communities by raising plantations on wastelands including community lands, degraded forests, other government wastelands and private marginal lands. Under government projects, funds were given only to government or aided institutions. However, it was observed that most of the degraded lands were under private controls and, therefore, the progress was slow.

Agro-Forestry

The third concept used for raising fuelwood through community participation has been under the agro-forestry programme, where farmers inter-crop cereal crops with tree plants on bunds, periphery of farmland or village land. This is mostly in an individual activity and very often the farmer does not find it financially lucrative.

Wasteland Afforestation

Under afforestation for reclaiming wastelands, a massive programme was developed by the National Wastelands Development Board with people's participation. The aims include the following:

(i) Establishment of decentralized nurseries and school nurseries;
(ii) Block Plantation especially on community land and lands owned by the scheduled castes and scheduled tribes and people living below the poverty line; and
(iii) Pasture development through people's institutions and involvment.

Energy Plantation

A major programme on bioenergy plantation was launched by the Government of India through the Ministry of Non- Conventional Energy Sources. Achievements under this programme include the following:

(i) Fuelwood Production and Improvement

The fuelwood production and improvement activity of the Biomass Programme envisages to increase productivity of fast growing short rotation fuelwood species suitable for plantation under a given set of soil and climatic conditions (initial pilot plantation under research centres has been observed to be up to 40 tonnes per hectare per year) and to develop suitable methodologies which *inter alia* is the main thrust of this activity. In order to achieve this objective, Biomass Research Centres have been set up in 11 different agro-climatic regions of the country till March, 1992. Biomass Research Centres are also being set up in the remaining four agro-climatic zones. Eighty different plant species have been identified so far. (Table 2.3 and 2.4).

TABLE 2.3: Agroclimatic Regionwise Location of Biomass Research Centres

S.No	Agro-climatic Regions		Biomass Research Centres
1.	Upper Gangetic Plains Regions	:	NBRI, Lucknow
2.	Southern Plateau and Hill Region	:	Madurai Kamraj Uni., Madurai
3.	Central Plateau and Hills Region	:	CTAE, Udaipur
4.	Western Himalayan Region	:	Garhwal University, Srinagar
5.	East Coast Plains and Hill Region	:	RPRC, Bhubneswar

6.	Western Plateau and Hills Region	:	Shivaji University, Kolhapur
7.	Trans Gangetic Plains Region	:	TERI, Gwalpahari (Haryana)
8.	Western Plateau and Ghat Region	:	University of Calicut
9.	Eastern Plateau and Hill Region	:	Ravishankar University, Raipur
10.	Lower Gangetic Plains Region	:	Viswabharti, Santiniketan
11.	Middle Gangetic Plains Region	:	Bihar University, Muzaffarpur
12.	Gujarat Plains/Hill Region	:	Proposed to be Set Up
13.	Western Dry Region	:	Proposed to be Set Up
14.	Eastern Himalayan Region	:	Proposed to be Set Up
15.	The Island Region	:	-

Nursery technologies have been developed and standardized for some of the fuelwood species to produce quality seedlings, and suitable species have been identified for plantation in some agro-climatic regions. Efficient strains of micro-organisms, capable of fixing atmospheric nitrogen, have been screened, tested and selected for field application.

(ii) Energy Plantation Activity

The main objective of the woody biomass energy plantation demonstration activity is now to test the performance of fast-growing fuelwood species and to screen the most potential species for plantation under a given set of agro-climatic conditions and to prepare the package of practices in close association with the concerned Biomass Research Centres. By March, 1992, 17,110 ha of land was under energy plantation.

(iii) Petrocrops

A survey sponsored by the government revealed that there are 386 different species of hydrocarbon plants in India. Out of these, 54 species emerged as potential species. Bio-crude from some of the hydrocarbon plants has successfully been extracted and its composition modified for use as a substitute of diesel/petrol. However, it has been observed that the extraction methodology developed is not economically viable for the present.

(iv) Liquid Fuels from Plant Sources

Efforts have been made to produce alcohol from ligno-cellulosic residues such as rice straw. An R&D pilot plant of 50-litre-per-day capacity for production for alcohol (ethanol) from rice straw has been set up.

TABLE 4: List of Plant Species Allocated to Biomass Research Centres for Preparation of Package of Practices

S.No.	Name of Centre	Name of Species
1.	NBRI, Lucknow	
		1. *Acacia nilotica*
		2. *Prosopis juliflora*
		3. *Dalbergia sissoo*
		4. *Poplar deltoides*
2.	MKU, Madurai	
		5. *Hardwickia binata*
		6. *Eucalyptus camaldulensis*
		7. *Gliricidia sepium*
3.	Garhwal University, Srinagar	
		8. *Acacia catechu*
		9. *Alnus nepalensis*
		10. *Bauhinia retusa*
		11. *Bauhinia variegata*
		12. *Quercus incana*
		13. *Ougeinia dalbergioides*
4.	CTAE, Udaipur	
		14. *Acacia tortilis*
		15. *Albizzia lebbek*
		16. *Prosopis cineraria*
		17. *Pithocelobium dulce*
5.	RPRC, Bhubaneswar	
		18. *Acacia auriculiformis*
		19. *Cassia siamea*
		20. *Casuarina equisetifolia*
		21. *Sesbania formosa*
6.	Shivaji University, Kolhapur	
		22. *Cassia fistula*
		23. *Eucalyptus hybrid*
		24. *Parkinsonia aculeata*
7.	TERI, Gwalpahari (Haryana)	
		25. *Haplophragma adenophyllum*
		26. *Terminalia arjuna*
		27. *Syzigium cuminni*
		28. *Melia azaderach*
8.	University of Calicut, Calicut	
		29. *Eucalyptus grandis*
		30. *Macaranga indica*
		31. *Terminalia catappa*
		32. *Xylia xylocarpa*
9.	Ravishankar University, Raipur	
		33. *Cleistanthus colinus*
		34. *Lagerstromia parviflora*
		35. *Peltophorum deltoides*

(v) Solid Fuels

Estimates indicate that the availability of agro-residues and agro-industrial wastes/byproducts alone is more than 246 million tonnes per annum which can be diverted for production of briquettes. Though the briquetting technology has already been commercialized in the country by private entrepreneurs, the existing technology suffers from various problems and as a consequence of this the rated capacities of the briquetting plants and machineries are not being achieved.

Possible Reduction in Fuelwood Demand by Higher Productivity and Efficient Utilization

Productivity Improvement

India's population in 1991 was 837 million, of whom 607 million were living in the rural areas. Energy consumption there is observed to be equivalent to 0.5 t/capita/year of bone-dried wood. That means the total energy consumption in rural India is equivalent to 303 mt/year of bone-dried wood. Actually only 55 per cent of this energy is contributed by fuelwood and the rest comes from cowdung and agro- residue. The average fuelwood yield for forest has been recorded as 0.5 t/ha/year. If all the 303 mt of fuelwood equivalent energy requirement was to be met from fuelwood alone, then plantation over 606 mha would be necessary.

Different experiments, conducted under projects funded by the Government of India, have indicated through energy plantation demonstrations an achievable average productivity of cultivation of 20 tonnes per hectare per year. In that case the total area required for meeting the fuelwood need would be only 15.15 million hectares (See Table 2.5). Yields recorded under 19 different bioenergy varieties in the research centres have shown productivity up to 36.80 tonnes per hectare per year (See Table 2.6). Thus, indicating possibility of raising fuelwood cultivation productivity to a level of about 40 tonnes per hectare per year.

TABLE 2.5: Fuelwood Production and Demand

1.	*Total population (1991)*	:	837 million
2.	*Total rural population (1991)*	:	607.1 million
3.	*Share of fuelwood in meeting non-commercial energy consumption of rural population*	:	65%
4.	*Fuelwood requirement per person*	:	0.5 t/person/yr

5.	*Total fuelwood requirement*	:	303 million tonnes/year
6.	*Average present productivity of fuelwood*	:	0.5 t/ha/yr
7.	*Total area required to meet the fuelwood need as per existing productivity*	:	606 million ha
8.	*Achievable productivity of energy plantations*	:	20 tonnes p/hr/ year
9.	*Total area required for meeting fuelwood needs*	:	15.15 million ha.

Note: Practically this figure refers to timber productivity from "Forest Trees".

TABLE 2.6: Biomass Yield of Selected Plant Species

S.No.	Name of Species	Age (Years)	Total Yield (t/ha)	Yield (t/ha/yr)	Source
1.	*Cassia siamea*	5	82.50	16.50	RPRC, Bhuba-neswar
2.	*Leucaena leucocephala*	5	62.50	12.50	-do-
3.	*Casurina equisetifolia*	5	31.00	6.20	-do-
4.	*Acacia auriculiformis*	5	60.60	12.12	-do-
5.	*Gliricidia sepium*	3.5	25.13	7.18	-do-
6.	*Trema orientalis*	3	54.00	18.00	-do-
7.	*Albizia stipulata*	5	101.40	20.28	Srinagar (U.P.)
8.	*Prunus cerasoides*	7	122.50	17.50	-do-
9.	*Acacia catechu*	6	116.64	19.44	-do-
10.	*Alnus nepalensis*	5	82.40	16.48	-do-
11.	*Albizia lebbek*	6	120.00	20.00	-do-
12.	*Ougenia dalbergioides*	6	97.68	16.28	-do-
13.	*Leucaena leucocephala (K8)*	3	110.40	36.80	MKU, Madurai
14.	*Albizia lebbek*	3	60.00	20.00	-do-
15.	*Hardwickia binnata*	3	100.50	33.50	-do-
16.	*Samania saman*	5	80.10	26.70	-do-
17.	*Cassia siamea*	3	62.10	20.70	-do-
18.	*Erythrina indica*	3	79.80	26.60	-do-
19.	*Prosopis juliflora*	8	96.49	12.06	NBRI, Lucknow
20.	*Terminalia arjuna*	6	42.00	7.00	-do-
21.	*Acacia nilotica*	8	59.50	7.43	-do-
22.	*Acacia auriculiformis*	8	54.54	6.82	-do-
23.	*Prosopis juliflora*	8.5	187.00	22.00	NARI, Phaltan

Efficient Utilization of Fuelwood, Agro-Residue and Waste

Besides production of biomass for energy, conversion of available agro-residues and agro-industrial waste into useful forms of energy is also important. It has been estimated that of the available 246 million tonnes agro-industrial waste the surplus agro-residue at 94.2 mt/year is available that can be briquetted and serve as a major source of energy.

Use in High Output Mode

It has also been estimated that instead of using biomass directly into furnaces and boilers, if they are converted into briquettes or are gasified, the input/output ratio substantially improves, from 1:0.98 in the case of direct use to 1:1.55 in the case of gasification.

Briquetting Technology can Replace 1mt of Furnace Oil by 2000 A.D.

Different technologies for briquetting have been developed and it has been shown that a one million tonnes replacement of furnace oil is possible in India by the use of 3.5 million tonnes of woody biomass. Land required for production of 3.5 million tonnes of fuelwood on a sustainable basis would be 0.175 million hectares at the rate of 20 tonnes per hectare per year. The present consumption of furnace oil in India is 10 million tonnes per annum. A national programme for conversion of 1 million tonnes of fuel oil by 2000, through efficient use of biomass, is being taken up.

The technological potential of briquetting in India has been shown to be sufficient enough to use all the 94 million tonnes of available agro-industrial waste per annum. This will require about 4,500 briquetting machines.

Gasification Technology for Process Heat

The next route for using biomass for process heat has been established through the gasification. A number of examples are available of improving the overall productivity of the industrial units and reducing their fuelwood consumption by using the gasification route.

Exciting Commercial Applications

Exciting examples are available of commercial applications of gaseous fuel from biomass. These examples are for direct gas application as well as power generation from the gas. Some of these, as developed by M/s ANKUR in the private sector in India are given below:

(i) Systems have been installed for *drying of tea* in South India where it has been clearly shown that indirect heat exchange (which leads to tremendous inefficiencies) is not required. Because of very high quality of the gas produced by systems and the fact that the gas is almost totally tar-free, combustion products can be directly mixed with air for drying of tea thereby providing efficiencies that are at least twice those in the conventional systems. Recorded savings are up to Rs. 100 per hour or about Rs. 1,000 per day. The investment in the gasifier is only Rs. 400,000 (Rs 30 equal to US $ 1).

(ii) A similar application taken up during the last few months has been in the *textile sector* where large quantities of biomass were burnt for cloth processing at different stages. Besides a considerable saving of biomass fuel, there are a number of other benefits in terms of better process control, elimination of cloth catching fire, etc.

(iii) A somewhat different application has been in Gujarat where a gasifier was installed more than three years ago so that *processing* of mangoes produced by the tribals could be carried out on their campus with reliable power availability. Now the entire campus (including all their PCs) is being run on a gasifier generation system.

(iv) Development of a 500 kw(e) gasifier system has been carried out under a Consortium Project with partial assistance coming in the form of a conditional grant from PACER (Programme for Accelerated Commercialization of Energy Research). The R&D work has led to the development and availability of a largest commercially available biomass gasifier system in the world. This system can be used for power generation, thermal applications or co-generation with the rated output being 500 kW of electrical power or 1.25 million cal/hr of thermal energy.

(v) Recently the company finalized a "sole-selling" arrangement for entire South India and Sri Lanka. This is the first effort where a manufacturer has been able to appoint a "selling agent" for his product.

Specific Issues

Land Availability Issues

Physical Tenure System Rehabilitation

India has a total geographic area of 328.72 million hectares with a forest cover of 63.91 mha and non-forest area of 256.94 mha. The total gross cropped area in 1987-88 was 172.88 mha.

It has been observed that the land holdings in India in 1985-86 were as given in Table 2.7.

Table 2.7: Proportion of Land Holdings

Area	No. of Farmers (in million)		Land-holding (in million ha)	
0-1 ha	56.147	(57.8%)	22.042	(13.4%)
1-2 ha	17.922	(18.4%)	25.508	(15.6%)
2-4 ha	13.252	(13.6%)	36.666	(22.3%)
4-10 ha	7.916 1	(8.2%)	47.144	(28.6%)
10 ha and above	1.918 1	(2.0%)	33.002	(20.1%)

Seventy-six per cent of the total farmers have land- holdings of up to 2 ha. In terms of area under cultivation these represent only 29.2 per cent of the land. These farmers arc involved only in subsistence farming. Productivity levels are low and mostly not profitable. These are located in drylands.

Cropped lands are such that 83 mha of the arable land is under marginal agricultural production which could be more productively used by re-vegetation with suitable species of trees, grass and legumes in association, especially in drylands and uplands.

Forest degraded area is 35.89 mha.

Non-forest degraded area is 93.69 mha in the following categories:

(i) Saline and Alkaline land - 7.17 mha
(ii) Wind-eroded area - 12.92 mha
(iii) Water-eroded area - 73.60 mha

Although, in general terms, the whole wasteland is reflected as can be covered under the biomass plantation for energy, only 7.17 mha of saline and alkaline lands can be utilized at a reasonable and affordable cost, that too after reclamation.

The *tenure system* of the land is such that in the case of most of the small land-holdings the farmers have only the right to sow and harvest and the land belongs to the government.

Rising inputs cost in India is forcing farmers with small land-holdings to consolidate either by grouping themselves or handing over lands to commercial organizations to achieve economic viability thus leading to a *rehabilitation process*. This will also result in a large number of people going out of agriculture and trying for an optional employment in urban areas. Such an exodus can be arrested only by providing a higher value of return for their production. The situation in drylands is likely to change at a faster pace than in other lands and, therefore, the initial attention for bioenergy plantations in India has to be focussed in the dryland areas and uplands and not in the totally degraded and marshy lands. Such an approach will on the one hand, provide a means to the farmer for a decent living on energy plantations, and, on the other, such plantations will meet all the energy needs and provide more local energy resource for an overall sustainable economic development.

Sustainable Feedstock Production and Use

Species, Inputs-Integrated Management, Field Handling and Transport—Fuelwood as a By-Product

A number of plant *species* have been identified depending on their suitability for different agro-climatic conditions. Eighty species have been mentioned in Table 2.8. Also, to suit development of species under different agro-climatic conditions, 11 different Biomass Research Centres (as mentioned in Table 2.3), are operating. These research centres provide continuous upgradation in technology packages and practices for a higher yield of the biomass species. The biomass yield of selected plant species in India has been given in Table 2.6 along with the location where such yields have been attained.

TABLE 2.8: List of Plant Species Selected for R&D Work at the Biomass Research Centres

S.No.	Name of Species	S.No.	Name of Species
1	*Acacia nilotica*	*9*	*A. stipulate*
2	*A. nilotica ssp cuprisiformis*	*10*	*A. procera*
3	*A. auriculiformis*	*11*	*Anogeissus pendula*
4	*A. catechu*	*12*	*A. latifolia*
5	*A. farnesiana*	*13*	*Alnus napalensis*
6	*A. tortilis*	*14*	*Azadirachta indica*
7	*A. moniliformis*	*15*	*Adhatoda vesica*
8	*Albizzia lebbek*	*16*	*Artocarpus heteropyllus*

17	*Alstonia chilansis*
18	*Adina wodive*
19	*Bauhinia retusa*
20	*B. variegata*
21	*Butea frondosa*
22	*Boehmeria regulosa*
23	*Rosewellia serrata*
24	*Casuarina equisitifolia*
25	*Cassia siamea*
26	*C. fistula*
27	*C. auriculatia*
28	*C. fistula*
29	*Callindra calothyrsus*
30	*Calophyllum enophyllum*
31	*Cleistanthus colinus*
32	*Dalbergia sissoo*
33	*Derris indica*
34	*D. sericea*
35	*Eucalyptus hybrid*
36	*E. citrodera*
37	*E. torreleyana*
38	*E. grandis*
39	*E. camaldulensis*
40	*E. globulus*
41	*Emblica officinalis*
42	*Eujenia spp.*
43	*Grewia spp.*
44	*Gliricidia sepium*
45	*Haplophragma adenophyllum*
46	*Hardwickia binata*
47	*Leucaena leuucocephala*
48	*Lagerstromia parviflora*
49	*Melia azaderach*
50	*Macaranga indica*
51	*Madhuca latifolia*
52	*Manihot utilissima*
53	*Olea glandulifera*
54	*Ougeinia dalbergioides*
55	*Prunus cerasoides*
56	*Pyrus pashia*
57	*Prosopis juliflora*
58	*P cineraria*
59	*Parkinsonia aculeata*
60	*Polyalthia longifolia*
61	*Populus deltoides*
62	*Pongamia pinnata*
63	*P. glabra*
64	*Pithocelobium dulce*
65	*Pterospermum accrifolium*
66	*Peltophorum deltoides*
67	*Phyllanthus emblica*
68	*Quercus glauca*
69	*Q. incana*
70	*Samania saman*
71	*Sinandra glauca*
72	*Sesbania formosa*
73	*Schleicher a trijuga*
74	*Spandias dulcis*
75	*Syzigium cuminni*
76	*Terminalia arjuna*
77	*T. tomentosa*
78	*T. Catappa*
79	*Trema orientalis*
80	*Xylia xylocarpa*

Each biomass research centre has been assigned the task of producing sapplings every year and then distribute them to the users after convincing them about their potential. Quality planting material is being produced in these research centres with a view to increasing productivity up to 40 tonnes per hectare per year. Pretreatment being given to plants ensures the survival rate of up to 90 per cent or above.

Inputs for such species are also a part of the "technology package" now being developed. Technical knowhow and package of practices are also being developed in these research centres and made available to user agencies. Training of user agencies and farmers has been an important component of ensuring effective utilization of inputs with the use of modern scientific methods.

Some of the problems faced in the field relate to lack of knowledge of the implementing agencies regarding the package of practices. Very often it becomes a hindrance in the speedy development and multiplication of plant material.

Field handling of the biomass has not caused any serious problem though transportation of biomass in terms of agro-residues is always a problem in the country. Large plantations, including the 1,000 ha plantation in Gujarat, are coming of age now and the actual problems relating to transportation will be experienced only with the forthcoming production.

Fuelwood production as a by-product from agro-forestry practices has been considered for quite some time but not practised so far. Eighty-three mha of marginal productivity lands provide an ample opportunity for growing fuelwood as a by-product. It will virtually cover all the land-holdings up to 4 ha covering about 90 per cent of the farmers (numbering about 88 million). Most of these farmers have seasonal employment or labour, and any activity relating to production of fuelwood as a by-product of agro-forestry at source. Since energy plantation may take away a substantial part of arable land, it will be necessary to introduce intensive agriculture practices to raise productivity of the remaining landmass for achieving desired crop levels.

Environment and Climate Change

Reduction of Green House, Carbon Sequestration, Biodiversity, Deforestation

The *green house gas reduction* is feasible more for the reason of higher efficient utilization of biomass and avoidance of fossil fuel consumption.

Higher growth of bioenergy plantations will ensure conservation of biodiversity and multiplication of the genetic base in the country. Many of the local varieties are getting extinct with an overall concentration on a few species. The genetic base of traditional fuelwood species is disappearing. Biomass Research Centres in India have identified 80 species suitable under different agro-climatic zones in the country as mentioned in Table 2.8. The information ensures preservation of plant biodiversity.

Energy plantations are normally not related to arresting deforestation. However, energy plantations have been raised mainly on the degraded arable agricultural lands. Their impact on arresting deforestation can be substantial if commercial scale re-vegetation of forests is taken up and alternative uses of firewood species can be commercially promoted. There are a few examples in India where eucalyptus is being used for making wooden boxes for packaging.

Socio-Economic Issues

Productivity and Conversion Costs — Scale of Application — Social Benefits — Externalities Subsidies — Financing

Production and conversion costs are directly related to the type of resource and location of use. However, some of the costs observed in India are as follows:

(i)	*Fuelwood Plantation*		
a.	Infrastructure costs on fuelwood plantation	-	Rs. 1,500 per tonne.
b.	Production cost	-	Rs. 100 per tonne per year
c.	Selling price	-	Rs. 500 per tonne (at the firm)
(ii)	*Briquetting*		
a.	Conversion cost of agro residue into briquettes	-	Rs. 1.5—2.0 per kg
b.	Selling price of briquettes	-	Rs. 3.0 per kg
(iii)	*Gasifier Application*		
a.	Electricity production from the gasifier	-	Rs. 1.50 per kwh
b.	Selling price of electricity	-	Rs. 2.00 per kwh

These costs relate to a specific site and may vary at other sites.

The above-mentioned costs have taken into consideration the capital investment on an annualized basis for the equipment for conversion.

As biogas and improved cookstove applications are not on a commercial scale, their cost of production has not been worked out.

Scale of application can be realized from the following figures of production capacities of equipment:

(i)	Briquetting Machine	-	2 tonnes per day
(ii)	Gasifier	-	500 kw (e)
(iii)	Biogas Plant	-	85 cu m

Social benefits of these devices are basically in terms of improvement of the economy and better living conditions because of availability of energy source. Direct economic and social benefits have not been worked out but their advantages in terms of health, especially in terms of reduced respiratory ailments, and improvement in industrial activities, have been a positive sign.

At present, energy plantation, biomass gasifier and similar production and conversion applications do not attract any *subsidies* from the government. In the past, the subsidies have proved an important instrument in neutralizing the cost advantage of conventional fossil sources of energy.

However, a pattern of subsidy exists on end-use devices for domestic application, especially the biogas plant, and the improved efficiency cookstove. This subsidy, for a private owner, is 33 per cent but for the scheduled castes and scheduled tribes this subsidy goes up to 50 per cent.

A new system of commercializing industry-oriented devices has been evolved under which the government intends to set up large demonstration units in each group of industries and then provide soft term loans to commercially viable technologies. The scheme of financing for different bioenergy equipment and devices is given in Table 2.9. The soft term loans are available at a rate of 10.30 per cent for users and at 15.5 per cent for manufactures of equipment.

Policies and Institutions

National, Regional, International

Different global scenarios and regional scenarios have been developed but not worked out in the form of a Plan of Action. The World Energy Congress 1992 came with the Global Perspectives on Renewable Energy Development, recommending certain policies and institutional arrangements. These are valid in the current context.

However, on a national and regional basis, certain issues have more pronounced importance depending on the type of fuels and applications in any specific country. In the case of India, furnace oil has been identified where consumption can be effectively reduced by the use of briquettes. However, it will require the government's specific order to ensure use of

TABLE 2.9: Details of Soft Term Loan for Biomass Devices.

S. No.	Sector	Interest Rate	'Promoter's Contribution	Term Loan	Ceiling on Term Loan	Repayment Maintenance Period	Period Total Repayment Period
1.	Biogas Generation	15.50% Rebate 0.5% for timely repayments.	25%	60% of Full loan amount	Rs.1.5 crores	2 years	8 years
2.	Biomass Utilization	10.30% Rebate 0.5% for timely repayments	25%	Full loan amount	—	2 years	10 years
3.	Co-generation for Power Generation	15.50% Rebate 0.5% for timely repayments	25%	40% of project cost	Rs.3.0 crores	3 years	10 years
4.	Manufacturing of High Efficiency Wood Burning Stoves	15.50% Rebate 0.5% for timely repayments	25%	Full loan amount	—	2 years	7 years
5.	Purchase of Biomass Gasifier for Power Generation	12.90% Rebate 0.5% for timely repayments.	25%	Full loan amount	—	2 years	6 years for new units . 5 years for existing units
6.	Manufacturing of (a) Briquetting Equipment (b) Biomass Gasifier	15.50% Rebate 0.5% for timely repayments	25%	Full loan amount	—	2 years	8 years

briquettes in certain types of industries. Similarly, gasifiers can be used for processing in altogether a different group of industries. Such applications will either replace fossil fuel consumption or else will reduce fuelwood consumption because of improvement in the efficiency of biomass that can be saved. But, it can be safely said that the fuelwood saved by improving efficiency of end-use equipment, based on biomass, can be utilized for replacing fossil fuels for other applications. This package can be evolved, over time, which would ensure reduction in fossil consumption and no increase in fuelwood consumption.

Setting up Regional Biomass Utilization Centres can help in providing training to local communities.

3

BIOENERGY POTENTIAL FROM JUTE, KENAF AND COIR RESIDUES

Jute, kenaf and coir are biodegradable, environment friendly, natural products. They have negligible adverse impact on environment, either during agricultural cultivation, conversion to jute and coir products in industry, transportation or usage of such products, and in their ultimate disposal after use. Jute and coir have also shown competitive superiority over synthetic materials due to their boidegradability. Jute and coir need low energy inputs for cultivation.

Jute is a major cash crop of Eastern India and is cultivated mainly in the States of West Bengal, Bihar, Assam, Orissa and Uttar Pradesh. Annual production of jute fibre was 1.2 mt. in 1991-92. Kenaf, an alternate to jute, is cultivated in the drier tracts of Andhra Pradesh, Karnataka, Maharashtra, Madhya Pradesh and parts of Bihar and West Bengal. Annual production of kenaf fibre was about 0.25 mt in 1991-92. Jute and kenaf are produced in 14 states, other states accounting for small quantities.

Coconut is a cash crop of South India and is cutivated mainly in Kerala, Tamil Nadu, Karnataka and Andhra Pradesh, though 13 states are reported to be producing coconut. The total production of white fibre coir was reported to be 1,27,000 tonnes and of brown fibre coir as 1,01,900 tonnes in 1992.93. Thus a total production of 2,28,900 tonnes is reported for 1992-93.

These are produced by the poor rural farmers who work on a

The Expert Group Meeting on Diversified Applications of Bio Resources such as Jute, Kenaf, Coir Sisal and Allied Fibres, organised by the Food and Agriculture Organisation (FAO), Bangalore, October, 5-7, 1994.

subsistence level. These crops are grown in rainfed areas. Irrigation has often resulted in shift from production of such fibre crops to paddy crop and other cash crops.

Over 0.5 million farmers, 0.4 million of them in Kerala, are employed in coconut production, harvesting and coir fibre separation. Eighty per cent of these farmers are women. Over 4 million farmers are involved in jute and kanaf production; 0.24 m are involveed in jute mills and 0.15 m are involved in other related works. Thus jute provides employment to 4.39 m overall employment in this sector is 4.89 m.

Types and Volumes of Fibre Residues and Wastes

In case of jute and kenaf, the residue after removal of the fiber during retting is only the sticks. These jute and kenaf sticks are comparable to any woody biomass and are a ready substitute for fuelwood. One kg of fibre results in 2.5 kg of sticks. Also during retting, the microbial action results in emission of methane enriched gas with characteristics similar to producer gas and suitable for burning, but too diffused for collection. Jute mill waste during spinning of the yarn is jute caddis. This can be mixed with cow-dung to produce biogas.

Waste from coir fibre extraction is called coir pith. Fibre is extracted in Kerala by the wet process of retting, and therefore, wet coconut pith is available. In case of fibre extracted in other states, coconut fibre is removed mechanically and the pith is available in the dry form. One kg of fibre leaves behind 2 kg of coir pith.

This pith can be used for making briquettes or for producing organic compost. The quality of compost is much superior to other manures as the water retention period is too long and nitrogen and potash contents are high.

Energy Composition of Wastes

Prof. P.D. Grover at the Indian Institute of Technology, has carried out characterization of jute sticks, jute dust, coconut shells and coir pith through elementary analysis and proximate analysis Table 3.1. Jute sticks sample was taken from Andhra Pradesh, in September 1990. Coconut shell and coir pith samples were taken from Kerala in 1992. Results of Elementary Analysis and Proximate Analysis are given in Tables 3.2 and 3.3.

Residue/Waste Availability

Volumes of production of different wastes/residues has been calculated on the basis of fibre production as mentioned above and is given in Table 3.1.

TABLE 3.1 : Volume of Residue/Waste Availability

	Waste	Fibre Production (mt)	Fibre: Residue Ratio	Residue/ Waste Production (mt)	Waste used as a traditional Energy Source mt (%)
i.	Jute & Kenaf Sticks	1.45 (1.20+0.25)	1.0:2.5	3.8	2.0 (50 %)
ii.	Coir Pith	0.23	1.0:2.0	0.46	Not Used
iii.	Jute Caddis	—	—	0.028	Not Used
iv.	Biogas During Retting	—	—	Not assessed	Not Used

Jute caddis is the waste left behind in mills.

Jute and kenaf sticks production in major jute and kenaf producing countries in Aisa is as follows:

(i) India	-	3.8 mt	(1991-92)
(ii) Bangladesh	-	1.8 mt	(1992)
(iii) China	-	1.8 mt	(1992)
(iv) Indonesia	-	70,000t	(1991)
(v) Thailand	-	0.3 mt	(1991)

TABLE 3.2: Elementary Analysis

	Carbon	Hydrogen	Nitrogen	Oxygen	Ash
Jute Sticks	47.8 %	6.6 %	1.8 %	42.6 %	1.2 %
Jute Dust	45.0 %	5.7 %	0.4 %	28.7 %	20.0 %
Coir Pith	NA	NA	NA	NA	NA
Coconut Shell	NA	NA	NA	NA	NA

TABLE 3.3: Proximate Analysis

	Calorific Value	Ash	Volatile Substances	Fixed Carbon
Jute Sticks	4402 kcal/kg	1.2 %	83.4 %	15.4 %
Jute Dust	4417 kcal/kg	19.9 %	69.2 %	15.9 %
Coir Pith	4146 kcal/kg	9.1 %	62.4 %	28.5 %
Coconut Shell	4249 kcal/kg	1.9 %	79.9 %	18.5 %

Present Pattern of Use

Of all the wastes listed above, only jute and kenaf sticks are presently utilized on a systematic and continuous basis as a source of energy.

Sticks are used as a domestic fuel, as replacement for fuelwood; and as thatching material. It has been estimated that of 3.8 mt availability in India only 75 per cent is utilized, 505 for burning in wood burning stoves in households and 25 per cent for thatching of roofs. The farmers and labourers involved in jute production are very poor and they utilize jute and kenaf sticks as a free fuel and thatching material. As a matter of fact the social system has so developed that farmers and labourers consider free availability of jute and kanaf sticks as their right, and a supplement to meagre wages.

Thus, about 2 mt of jute and kenaf sticks are utilized as replacement for fuelwood and 1 mt are utilized as thatching material.

Jute caddis is presently not put to any use.

Gases emitted during retting also do not find any use.

Coir pith is burnt for domestic use after drying, but no estimate has been made for their pattern of use.

Economics of Use of These Residues/Wastes

Jute and Kenaf are not traded commodities and therefore their market value and financial implication cannot be quantified. Present use pattern is such that jute and kenaf sticks replace fuelwood in household cooking and thatching material in the housebuiliding sector. Its specific implication is the social security in terms of availability of fuel for cooking and thatching material for shelter for the poor farmers and labour involved in jute and kenaf farming. The sticks are available free to the farmers, and therefore, saves them money they would have spent on procuring fuelwood.

The economics can best be assessed in terms of the cost of materials being replaced. To work out replacement of fuelwood, the calorific value is taken as the criterion. As the calorific value is taken as the criterion. As the calorific values of fuelwood of jute and kenaf sticks are close (as assessed by Prof. P.D. Grover) the useful energy from 1 kg of fuelwood and 1 kg of jute/kenaf sticks is assumed to be same. Most of the jute farming areas are deficient in fuelwood, and people may have to spend Rs.1.0 to 1.50 per kg.

Thus, use of jute and kenaf sticks at 2 mt per year for fuelwood replacement reflects a saving of 2 mt of fuelwood valued at Rs.2000 m/ per year. The additional environmental advantage is that it reduces the pressure of deforestation.

Also use of 1 mt per year of jute and kenaf sticks as thatching material will replace other thatching material worth atleast Rs. 500 m per year.

As such the jute and kenaf residues are utilized to a substitution value equivalent to Rs. 2,500 m per year,

All these wastes are burnt in traditional equipment and devices and therefore do not incur any extra cost on this account.

Environmental Advantages and Disadvantages of the Present Pattern of Use

FAO's International Consultation on "Jute and the Environment" held at the Hague (The Netherlands) during October 1993 emphasised on the need for environmental impact assessment of jute and jute products for a number of purposes: ecolabelling, advertising and product improvement. The assessment processes had to include four steps: goal definition, data inventory (or collection), impact assessment and improvement analysis. The Consultation suggested that careful consideration of processes, products and wastes for the entire life-cycle is required for an objective and complete analysis. With regard to jute, there is a need for setting up a validated inventory including all steps in the life cycle. For example, more detailed information on fertilizer and pesticide use in jute and kenaf cultivation, as well as on the impact of the retting process is needed.

In its recommendations on R&D activities, the Consultation included that research is urgently needed to demonstrate and establish/promote the use of jute in activities such as composting and generating of biomass energy.

The Consultation obviously considered appropriate to have the authentic data to work out life cycle generations and to internalise various costs to arrive at economics of jute production. The Consultation also emphasized on impact of retting and generating biomass energy from jute and kenaf wastes. These steps have impact on environment and economics.

Jute, kenaf and coir are natural fibres and are biodegradable. They are not known to produce any harmful toxic products. M. Braungart, et al in Jute and Poly propylene - Comparative Environment Impact Assessment published by EPEA (Germany) in 1992 has indicated that jute and kenaf remove a net about 1.2 t of carbon dioxide from the atmosphere for each tonne of fibre produced.

In about 4 months' time jute crop produce about 3.0 t/ha of fibre and about 7.0 t/ha of sticks. Besides carbon dioxide absorption, the jute sticks meet the need of cooking fuel of the rural people in J & K in jute and kenaf growing areas, chronically short of other sources of fuelwood and avoid deforestation.

Jute sticks are also yielding particle boards and saving valuable timber woods. These sticks are also being considered for paper pulp. Leaves and other vegetative fraction, and retting wastes are valued as organic manure.

Coir pith is not having as many uses and advantages as jute and kenaf sticks. However, its use as a source of organic manure are certainly environment friendly.

Anwar Alam in Jute Retting and Environment, published in Prceedings of IJO Workshop 1993 has reported "emission of methane, hydrogen sulphide and carbon dioxide during retting of jute and kenaf is reflected as adversely affecting the environment". Retting is a phenomenon during which microbes present in water or surface of biomass, get active, digesting the tissues and substances adhering and holding the fibre bundles. In case of jute and kenaf the retting may be completed in 7—30 days depending upon the fibre crop, temperature, pH, microbes present etc. Thus, the gas emission is for a small time period. E.J. Zachariah, et al in "Retting Gas—A Biomass Energy Source" in Integrated Renewable Energy for Rural Development have reported emission of similar gases during retting coconut husk in the backwaters of Kerala. Unlike jute and kenaf, coconut retting continues for 8-10 months in a year, except for monsoon period. This makes possible to collect these gases for regular supply at an economic cost and limited investment. As of present these gases are just released to atmosphere. As the gas supply can be maintained throughout the year, the fuel gas can be used in any ancillary industry associated with coir production. It may also be possible to use this gas for cooking purposes by rural people.

Anwar Alam has suggested development of retting-cum-biogas digester that might permit retting under controlled conditions (thereby avoiding release of gases), yielding high quality fibre and two useful major by-products, viz, biogas a source of energy better than fuelwood, and liquid and solid wastes valued as manure.

Thus, on the environmental advantage side, besides being biodegradable, production of fibre crops works as carbon sink; and utilization of residues and wastes as fuel avoids deforestation and retains heat sinks of the natural forests. Its adverse environmental impact is due to release of greenhouse gas, viz, methane during fibre separation by retting. Technology development can lead to these gases being used as a source of energy.

Besides the possible environmental advantages of hard fibre crop production, efficiency improvement in residue use can also bring in additional environmental advantage.

Evaluation of Potential of these Wastes as Fuel for Rural Households and Agro-Industries

Potential of these residues for rural households and agro-industries can be projected only after considering the present use pattern and the possibility of efficient utilization, so that the energy needs of present consuming communities is not disturbed. Following possibilities may be considered:

(i) Higher production of biomass and corresponding increase in residue/waste in case of jute, kenaf and coconut;

(ii) More effective utilization of biomass for fibre extraction resulting in higher production of residue/waste without increasing biomass plantation production. In case of coir the present utilization is only 30 per cent of coconut husks for coir fibre extraction; and is proposed to be raised to 50 per cent thereby resulting in correspondingly higher pith availability; and

(iii) Efficient utilization of residue/waste, for example, jute and kenaf sticks burning in high efficiency wood stoves; and conversion of residue/waste into densified fuel through briquetting, or gaseous fuel through gasification.

As jute related farming work is only for a period of about 4 months, there is substantial scope for starting rural industries by providing small quantities of energy either in the form of process heat or electricity. Quantities as low as 5-10 kw can also bring around a qualitative change.

The rural communities have to be explained on the need for efficient utilization of jute, kenaf and coir residues as that will generate income opportunities for the remaining part of the year. Entrepreneurial development for rural industries — either traditional or agro-based shall have to be initiated with the involvement of enterprising industrialists. A concept has to be developed and nurtured. The initial phase may have the situation of chicken or egg. We need both - egg after preserving the chicken. The industrialists, big or small, shall be attracted only when they are assured of regular supply of source of energy (i.e. jute, kenaf and coir residue) and proven technology for its conversion into the required use form.

To achieve this objective it is necessary to offer proven, site specific, low cost technologies so that an attractive return is assured to the industrialists. Such small scale rural systems need to be backed up with security and quality of energy supply coupled with lower running costs. This calls for devising cost-effective ways of collecting, handling, transporting and storing the jute and kenaf sticks and coir pith and establishing a regular distribution network of the resultant product. Thus, the systems shall have to be efficient for making economic sense. The social and developmental benefits derived from the indigenous provision of energy and employment shall outweigh all other considerations.

The overall potential, therefore, shall have to be assessed in the light of these facts.

By introduction of high efficiency wood stoves, the jute and kenaf sticks utilization for cooking energy can be reduced from the present reported 2 mt per year to about 1.3 mt per year, thereby saving 0.7 mt of sticks residue.

All this can then be briquetted for use in local rural industries. A number of briquetting units can be set up in the jute, kenaf and coconut growing areas. (Part of this available biomass can also be gasified for electricity).

Projected higher utilization of coconut husks for coir making, from 30 per cent of coconut husks to 50 per cent, will increase availability of coir pith from 0.46 mt to 0.69 mt per year. Moreover, of 25 per cent remains unavailable in future for energy purposes, 0.55 mt per year of coir pith would be available.

The most attractive route is briquetting. These briquettes will be found useful by rural industries. The briquettes can be used either for direct burning or for gasification, depending upon the requirement of the rural industry.

In addition, development of retting-cum-biogas digesters can result in systematic collection of biogas produced during retting. Table 3.4 gives sector-wise potential applications:

TABLE 3.4: Sector-wise Potential Applications.

Sector	J&K Sticks	Coir Pith	Remarks
Household Sector	1.3 mt	—	Coupled with use of high efficiency wood stoves so that the effective cooking energy required by farmers is not affected.
Rural/Agro Industries Sector	0.7 mt	0.55 mt	J&K sticks to be used through briquetting and gasification Coir pith to be used through briquetting and may be subsequent gasification.

Options for Conversion of Residue/Wastes into Biofuel

Different technologies of conversion not in use currently are discussed below.

Following distinct possibilities of use of residues from jute, kenaf and coir fibres exist for their use as a source of energy for rural households and rural industries in areas around the raw material farms (besides the present use by direct burning):

(i) Briquetting of jute and kenaf sticks, and coir pith after grinding (and mixing with another suitable material like saw dust, if necessary) and use of resulting briquettes by rural industry (or for gasification);

(ii) Gasification of jute and kenaf sticks directly, and use of resulting producer gas used either directly for process heat or for electricity generation; and

(iii) Biogas production during retting of jute, kenaf and coconut husk in specially designed biogas-cum-retting digesters and by mixing jute cuttings and jute caddis with cow dung, in regular digesters.

Current fibre extraction practices and use pattern of jute sticks shall/ be dependent on the technological development.

This part describes some work already initiated for Technology Development and options for adoption of technologies developed for other raw materials for generation of energy from biomass:

Briquetting

Fibre wastes are normally voluminous, scattered, having low thermal efficiency (by volume) and liberate smoke. To overcome these adverse effects densification by compacting can be practiced. This practice is called briquetting.

Two possible routes of briquetting exist:

(i) Binderless Briquetting
(ii) Briquetting with a Binder.

For the wastes, under consideration, the binderless technology would be more suited. No binders are added in this technology, as lignin contained in the residue comes out as a fluid at a temperature of 150-200°C and acts as a binder. Two options have been developed and tested in India for various agro wastes including coir pith. No one has tried on jute and kenaf sticks. Jute Technological Research Institute, JTRL, and the Indian Jute Mills Association, IJMA, are seriously considering these options.

A number of manufacturers in India have started developing die punch machines. It has three stages, viz, crushing meterial to obtain required ideal particle size; selective drying to reduce moisture content to 10-15 per cent; and compacting of waste into briquettes.

In case of machines manufactured by Solar Science Consultancy, solid briquettes of 60 mm diameter are produced at a rate of 500 kg/hr. One of the users of the machine, Col. Devdas at Tirunelveli, has regularly produced briquettes from coir pith having moisture content of 60-70 per cent. The coir pith is mixed with saw dust in the ratio of 70 per cent to 30 per cent. Daily production of 6 to 7 tonnes/day has been achieved. Such briquettes have found a ready market at Rs. 1,600 per tonne as against the production cost of less than Rs. 1000 per tonne. The machine has been in operation for over 2 years. Total cost of machine with installation is about Rs. 1.0 m.

In another case, Indian Institute of Technology, Delhi has set up a pilot plant using Binderless Technology, utilizing Screw Compacting Mechanism. In this case briquettes of 55 mm outer size with a central bore of 15 mm are produced at a rate of 400 kg per hour. The external surface is carbonised at 300°C, using a heater at the compacting stage. Due to carbonising , the outer surface is protected against moisture and helps in burning. The punched hole in the centre helps in better air supply. A number of waste materials have been tried though jute, kenaf sticks and coir pith have so far not been tried.

Locally delivered cost of briquettes at about Rs.1600 per tonne makes it extremely attractive in fuelwood deficient areas. The cost of production is in the range of Rs.750 to Rs.1000 per tonne only.

Total cost of machine with installation will be about Rs.1.5 m.

Gasification

Among the various technologies utilizing biomass, the gasification route offers good promise as it provides process heat as well as the gas for power generation, making it ideally suitable for applications in rural areas.

A number of designs of gasifiers have been developed and used in India. However, as jute and kenaf sticks are to be used, the gasifier developed by the Chemical Engineering Department, Indian Institute of Technology, Delhi and run on lantana sticks, growing as a weed in the hilly regions, will be suitable. These sticks have a marked resemblance with jute and kenaf sticks and will not require any special arrangement.

This down draft gasifier is ideally suited for agro-forestry residues having ash contents less than 4-5 per cent on dry weight basis. (jute and kenaf sticks have 1.2 per cent ash content).

Biomass material is partially pyrolysed (carbonised) in the drum charring unit. In this process of pyrolysis (originally used to produce charcoal) the solid biomass is heated in the absence of oxygen or air to form char (charcoal), that comprises about 30 per cent by weight of the feed stock and contains 42 per cent of its calorific content. It is sized and fed to the gasifier where heating is initiated using the hand blower. The producer gas is cooled and cleaned in the hydro-separator before it enters the surge tank. The gas is then either burnt directly or fed to the engine-pump/alternator set as per requirement.

By-product from this is powdered char which can be manually briquetted to give smokeless domestic fuel.

Since the biomass is first charred before gasification, the problem of tar formation is overcome.

It has been estimated that the residue consumption shall be about 1.5 kg (of char) per kwh.

The current cost projections of the gasifier system indicates that the cost of gasifier will be Rs.35,000 and that of the generator will be additional Rs.65,000.

Since only 10 such systems of 5 H.P. have been run so far it may be necessary to do some more industrial R&D utilizing the fibre waste.

Biogas

Biogas is produced when jute and kenaf stems and ribbons, and coconut husks are subjected to retting.

In case of jute and kenaf, when stems or their ribbons are put to retting, biogas is produced in about 1 week and continues till the end. The retting period varies between 7 days and 30 days depending upon the temperature. Retention time for biogas production from cattle dung and pulverized crop residues may be about 20-30 days if the ambient temperature is around 35°C. Maximum temperatures for retting are 30-35°C at which stem retting may be completed in 10-15 days. Though both the processes are anaerobic, the requirements of the two are different. The retting period normally is about 2-3 months a year. That has been a major consideration why scientists did not think in terms of biogas collection process. The retting gas has been observed to contain upto 62-65 per cent methane.

In case of retting of coconut husks, large bundles consisting of about 3000 to 5000 husks are made and immersed in water for long periods ranging from eight to twelve months. In one of the analysis conducted in a rettery near Thiruvananthapuram, the rettery gas was observed to contain 32 per cent methane and 14 per cent carbondioxide.

Since the gas is produced round-the-year it is possible to design and develop gas collection mechanism at an affordable cost. Anwar Alam has suggested technical possibility of developing retting-cum-biogas production digesters, operating in a series, a battery of 3-5 digesters, which can produce and supply biogas at reasonably constant rate. Alam has cautioned that one has to have complete understanding of the end-point of retting when a particular digester could be loaded again, with some supplemental nitrogen in the form of urea or other nitrogen rich locally available resource, and digested till biogas production tapers to too low levels. Hopefully by that time sticks will be so soft that they could be easily pulverised and converted into valuable compost manure. A retting-cum-biogas plant, with manure as a by-product, may be an attractive proposition. That will be environmentally sound and effective compared to prevailing practices. The system may have to operate on multiple feed stocks as the year round availability of jute and kenaf raw materials cannot be anticipated.

Although some laboratory work has been done on biogas plants working on jute/kenaf stems or ribbons there is no commercial size plant in operation. some laboratory level work has been done in India at Jute Technologies Research Laboratory, (JTRL), and Indian Jute Industry Research Association, (IJIRA).

IJIRA, in its experiment, is using a floating dome type digester producing biogas from jute cuttings mixed with cattle dung slurry.

JTRL, in its experiment, used batch type of digesters. JTRL based its programme of biogas production on jute caddis.

Measures for Commercialization and Marketing of these Fuels

Following recommendations are framed to promote better 'commercialization and marketing' of jute, kenaf and coir fibre residue based fuels:

(i) A micro-level resource assessment (availability) linked area development plan for jute, kenaf and coconut growing areas needs to be prepared to promote decentralized development through local/rural industrialization using process heat and electrical energy from jute, kenaf and coir fibre wastes.

(ii) Jute and Coir Industry should be encouraged to promote/develop captive plantations so that the waste is managed and utilized more efficiently.

(iii) An intensive awareness programme needs to be launched for the farming community informing them about the alternative uses of jute, kenaf and coir fibre wastes as a source of enrgy that can create more income generation and employment opportunities and provide better quality of life. The focus should be on the possibility of creation of income generation opportunities during 7-8 months in the year when jute related activities stop.

(iv) Demostration of improved woodstoves, gasifiers and briquetting plants should be conducted in all major fibre producing areas. This will build confidence about mature technologies in users and enterprising persons willing to start rural/agro industries.

(v) Intensive research needs to be undertaken in retting-cum-biogas plants so that the gases emitted during retting are gainfully utilized and also environmental pollution is avoided.

(vi) Awareness should be created in gasifier and briquetting machine manufacturers about energy potential of the available jute, kenaf and coir fibre wastes and the emerging markets. These manufacturers should be enthused to undertake marketing of their products.

(vii) FAO may develop a Project on "Awareness Creation, Demonstration and R&D Aimed at Commercialisation of Briquetting, Gasifier and Biogas technologies utilizing Jute, Kenaf and Coir Fibre Wastes".

The Project should be a collaborative effort amongst FAO, Governments, Jute/Coir Mill owners and Gasifier and Briquetting machine manufacturers to build confidence in the technology and initiate action with mutual sharing of investments.

The ultimate objective shall be utilization of jute, kenaf and coir fibre wastes on a self-sustaining basis under bankable schemes.

4

GLOBAL PERSPECTIVE IN BIOENERGY USE

Biomass, in all forms, supplies about 14 per cent of the world's primary energy. In the developing world, fuelwood provides a source of energy for basic subsistence, i.e. for cooking; and in the developed world, fuelwood is used either for power generation or process heat. In view of specific nature of fuelwood, it is presently the largest energy source in the developing countries where it produces 35 per cent of the primary energy (90 per cent in some areas). Since nine-tenths of the world population may reside in the developing countries, by the middle of the 21st century, biomass energy is likely to remain an important global energy source. The 15th Congress of the World Energy Conference held at Madrid in September, 1992, emphasized this aspect.

Present Pattern of Global Energy Use

Over the past 30 years, the world's energy requirement has risen considerably. In 1960, the world used 3.3 Gtoe (giga tonnes of oil equivalent) of energy; and in 1990, 8.7 Gtoe, an increase of 163 per cent or an average increase of 3.3 per cent per annum.

Energy use per capita is even more varied. At present the majority of the world's population uses less than 1.2 toe (tonnes of oil equivalent) of energy per capita (indeed many use less than a third), the energy rich (in North America) use more than 7 toe per head. Under the commercial energy, the per capita energy consumption in India is still one of the lowest in the world. It recorded 123.5 kgoe (kilogrammes of oil equivalent)

International Workshop on Renewable Energy Project Planning and Management organised by UNDP and IIT, Delhi, on March 23-25, 1993, New Delhi.

during 1990-91. However, if non-commercial energy consumption is also included, the per capita consumption goes up to 265.4 kgoe. The per capita electricity consumption in India recorded 253 kwh. However, the growth in demand has been fastest in the developing countries. In the past decade, it has risen in developing countries by 49 per cent, and in the developed countries by only 14 per cent. Although most of the energy is still consumed by the developed industrialised countries, the proportion accounted for by the developing countries can be expected to continue to increase.

Global Energy Perspectives - WEC Report: "Energy for Tomorrow's World"

The WEC Commission has developed three Global Energy Cases — each representing a different assumption in terms of economic development, energy efficiency and environmental impact.

(i) The first case is referred as the Reference Case (REF). It is in essence the same as developed by the WEC for the Montreal Congress in 1989, updated by recent experience, and considering "business-as-usual" scenario;

(ii) The second case is the Enhanced Economic Development Case (EED) which assumes a somewhat higher rate of economic growth in the developing countries than the Reference Case but with somewhat reduced energy efficiency; and

(iii) The third case is of the Ecologically Driven Case (ED) which assumes the same economic growth as in the Reference Case, but with a sharper improvement in energy efficiency.

In each of the three cases the critically important projection of population growth is assumed to be the same, and is taken from the current U.N. Projection — implying an explosive increase in world population from 5.3 billion in 1990 to 8.1 billion in 2020. The projections of the World Energy Demand up to the year 2020 corresponding to these cases are presented in Table 4.1:

TABLE 4.1 : Future Energy Demand — Three Cases.
(Gigatonnes of Oil Equivalent, Gtoe)

1990	2020		
	REF	EED	ED
8.7	13.3	17.2	11.2

Table 4.2 reflects the possible Global Energy Supply Mix for the three cases in the year 2020 as compared to the corresponding energy mix in 1990.

TABLE 4.2 : Energy Mix in Global Fuel Use.
(Gigatonne of Oil Equivalent, Gtoe)

Energy Form	1990	2020		
		REF	EED	ED
Coal	2.3	3.2	4.8	2.1
Oil	2.8	3.7	4.6	2.7
Gas	1.7	2.8	3.5	2.3
Nuclear	0.4	0.8	1.0	0.7
Hydro	0.5	1.0	1.2	0.9
New Renewable	0.2	0.5	0.8	1.5
Traditional* Biomass	0.8	1.3	1.2	1.0
Total	8.7	13.3	17.2@	11.2

* Only firewood, agro residues and cowdung and does not include biomass recoverable by new technologies.

@ Rounding of the figures show a deviation of 0.1 in the total.

Renewable Energy Resources: Opportunities and Constraints — 1990-2020

The contribution of new and renewable sources of energy in the global context is considered small compared to the commercial sources of energy. As a group, they are difficult to compare to the conventional energy resources, since they are widely distributed and relatively diffuse, only some of the technologies have been developed to commercial status, and ask for somewhat different compromises from the end-user. Most are or can be made environmentally benign and the supplies are vast, but they are perceived as expensive and unreliable. There is a need to apply strict environmental standards to renewable energy development, as part of a consistent process towards total energy provision and use, but costs can be confidently expected to fall and reliability to rise.

The WEC Working Group on Renewable Energy Resources also projected future needs as follows:

Projected Use Scenarios

The Renewable Energy Report was developed in two different scenarios to study the use pattern of renewable sources of energy in the global context. These are as follows:

Scenario 1 : Continuation of current policies without special emphasis on renewable energy use.

Scenario 2 : Renewable energy supplies as a part of the global ecology initiative and intended to represent higher penetration of global market than is reasonable by 2020.

The philosophy of these two scenarios is the realization for giving impetus to renewable energy systems in the same fashion as was given to the conventional energy sources. A major concern obviously was the decision about its use pattern. Energy decisions have to address to long-term impacts otherwise the renewable energy sources will make little headway in the energy market. As such, an examination of the constraints was the starting point. Various sources considered include: solar, wind, geothermal, ocean and bio-mass. Small hydro is also covered under the similar category.

The current use of renewable energy resources is dominated by biomass which is a major energy supply in the developing countries of Asia and Africa with some additional contribution in North America and Latin America. When large-scale hydro use is included, the renewable resources contribute about 18 per cent of mankind's total energy use. Without traditional biomass and large hydro, the current contribution of all other renewables is about 1.9 per cent. To build on such a small base and make a significant contribution in three decades is a major challenge, considering that the nuclear energy achieved only a 5 per cent share in 30 years of commercial development with sustained government supported R&D.

The first scenario referred as Business-as-Usual, which is based on a slow increase in price of conventional fossil fuels, continued developoment of renewable energy technologies at a modestly increasing pace as their economic market opportunities expand, increased pressure, over time, on energy efficiency and environmental protection, and sustained economic development. The total contribution by 2020 may be higher than previously estimated by the WEC, with a growth of 3.4 times compared to 1990, but still a relatively modest contribution of about 4 per cent in the overall. When traditional biomass and large hydro are included, the total renewable contribution will represent about 1.8 times the current level and will account for an estimated 21 per cent of the global total energy supply. With the exception of traditional biomass, this penetration is primarily in the power sector, with very little impact on the transport sector, although in some countries, notably Brazil, Zimbawe, Kenya and Cote d'Ivoire, biomass derived fuels reached national importance. Table 4.3 gives the growth pattern under this scenario.

Renewables are expected to compete primarily with fossil fuels, but hybrid systems that share the advantages of both renewable and fossil fuels hold near term promise.

In the second scenario, a faster and more extensive penetration of the renewable into the energy market has been forecast. This could happen through the recognition that the external costs of renewable energy are much lower than those for conventional energy use when consistently high environmental standards are applied to all energy forms. These external costs are not being cosidered in micro-economic decision-making today and if applied in consumption with greater energy efficiency and other positive measures, it is clear that the penetration can be much higher. Under this scenario the contribution of renewables can increase by 10 times to 1,533 Mtoe which will be about 15.3 per cent of the total consumption. If traditional biomass and large hydro are also included then the total contribution can increase to 3,450 Mtoe which will be slightly over 30 per cent of the total energy consumption. Table 4.4 gives growth under this scenario.

Projection of Global Contribution of Biomass

Biomass provides the possibility of conversion into liquid fuels, solid fuels and gaseous fuels. This versatility opens new opportunities for its use for different applications. This will make it significantly important for use in the 21st century. It will also continue to meet its important traditional roles, i.e. cooking, heat, light, construction, etc., in the rural and urban sectors of the developing countries. Biomass is a flexible feedstock capable of conversion into an appropriate form which can substitute for fossil-

Table 4.3

Summary of Regional and Global Projections for Current Policies Scenario

Values are MTOE	Solar	Wind	Geo-thermal	Biomass (3) Trad.	Biomass (3) Modern	Ocean	Small Hydro	Large Hydro (1)	Renew. Total	% of Global (1)	"New" Total (2)	% of Global (1)
1990												
North America	3	1	6	38	19	0	4	127	198	2.3%	33	0.4%
Latin America	1	0	2	125	46	0	1	86	261	3.0%	50	0.6%
Western Europe	1	0	1	20	10	0	7	98	137	1.6%	19	0.2%
NIS & East Europe	2	0	0	30	10	0	2	53	97	1.1%	14	0.2%
MidEast/No. Africa	1	0	0	21	0	0	0	4	26	0.3%	1	0.0%
Sub-Saharan Africa	1	0	0	141	5	0	0	8	155	1.8%	6	0.1%
Pacific/China	1	0	3	351	23	0	4	70	452	5.2%	31	0.4%
Central/South Asia	2	0	0	204	8	0	0	20	234	2.7%	10	0.1%
TOTAL	12	1	12	930	121	0	18	466	1560	17.9%	164	1.9%
	0.8%	0.1%	0.8%	59.6%	7.8%	0.0%	1.2%	29.9%	100.0%			
2000												
North America	6	3	10	40	26	0	5	143	233	2.3%	50	0.5%
Latin America	1	0	3	145	54	0	2	126	331	3.2%	60	0.6%
Western Europe	1	1	4	20	14	0	8	119	167	1.6%	28	0.3%
NIS & East Europe	3	0	2	32	12	0	3	54	106	1.0%	20	0.2%
MidEast/No. Africa	2	0	0	28	0	0	0	7	37	0.4%	2	0.0%
Sub-Saharan Africa	1	0	0	198	7	0	0	12	218	2.1%	8	0.1%
Pacific/China	3	1	7	377	29	0	5	112	534	5.2%	45	0.4%
Central/South Asia	2	1	0	240	10	0	0	29	282	2.8%	13	0.1%
TOTAL	19	6	26	1080	152	0	23	602	1909	18.7%	227	2.2%
	1.0%	0.3%	1.4%	56.6%	8.0%	0.0%	1.2%	31.5%	100.0%			
2010												
North America	13	14	14	43	40	0	6	154	284	2.4%	87	0.7%
Latin America	5	0	4	163	63	1	2	204	443	3.7%	75	0.6%
Western Europe	4	7	5	20	19	0	9	130	195	1.7%	45	0.4%
NIS & East Europe	6	1	3	34	18	1	5	58	126	1.1%	34	0.3%
MidEast/No. Africa	4	0	0	32	0	0	0	11	47	0.4%	4	0.0%
Sub-Saharan Africa	2	0	0	250	10	0	0	19	282	2.4%	12	0.1%
Pacific/China	7	4	10	398	36	1	8	145	610	5.2%	66	0.6%
Central/South Asia	5	2	0	268	12	0	1	47	335	2.8%	20	0.2%
TOTAL	46	29	37	1208	198	3	31	769	2321	19.7%	344	2.9%
	2.0%	1.2%	1.6%	52.0%	8.5%	0.1%	1.3%	33.1%	100.0%			
2020												
North America	31	42	21	46	55	1	6	167	369	2.7%	156	1.1%
Latin America	14	1	6	179	72	3	4	304	583	4.3%	100	0.7%
Western Europe	10	18	8	20	24	4	10	145	239	1.8%	74	0.5%
NIS & East Europe	14	4	4	36	23	1	7	63	152	1.1%	53	0.4%
MidEast/No. Africa	10	1	0	38	0	0	0	16	65	0.5%	11	0.1%
Sub-Saharan Africa	5	1	0	299	12	0	0	29	346	2.5%	18	0.1%
Pacific/China	23	13	15	414	43	5	12	188	713	5.2%	111	0.8%
Central/South Asia	15	5	0	291	14	0	1	70	396	2.9%	35	0.3%
TOTAL	122	85	54	1323	243	14	40	982	2864	21.0%	559	4.1%
	4.3%	3.0%	1.9%	46.2%	8.5%	0.5%	1.4%	34.3%	100.0%			

1) Based on new WEC M* scenario (with interpolation where necessary) – Global Total in 2020 is 13 606 MTOE
2) New equals Renewable Total less Traditional Biomass and Large Hydro
3) Biomass is divided into Traditional and Modern Categories

based fuels at a relatively small scale of investments. Many benefits may additionally accrue from significantly increasing the use of modern forms of biomass energy. These include the reduction of carbon dioxide emissions into the atmosphere if advanced biomass conversion technologies are used to substitute for fossil fuel- based technologies. Further more, the adoption of advanced biomass energy conversion systems may well be less costly than equivalent advanced fossil fuel-based systems.

The biomass energy use potential in 2020 is estimated at 1,566 Gtoe under the current policies scenario and 1,711 Gtoe under the ecologically driven scenario. J. Wood and D.Hall in their study 'Bioenergy for Development' —FAO Environment and Energy Paper 13, 1994 have estimated that currently biomass provides about 14 per cent of the global primary energy consumption and could in the future, theoretically, supply the equivalent of about thrice the present global energy demand. This scenario allows for increases in the cropland area for food production by the year 2025. More recently, the perception of biomass energy has changed in a number of industrialized countries. This has led to biomass gaining a growing and significant share of the primary energy sector in the USA, Sweden and Austria (4, 16 and 10 per cent of the primary energy respectively). Biomass is increasingly viewed as an environmentally and socially advantageous source of energy.

Biomass could become a central part of future sustainable energy supply, through both economic and technical feasibility. Such a developmental approach has been demonstrated by Johannson et al (1992) in their *Renewable Intensive Global Energy Scenario* (RIGES). It is demonstrated that there is a possibility of providing energy for growth and development at no extra cost compared to conventional fossil-based systems. A reduction in the global carbon dioxide emissions would occur as a result of such an increase in biomass-based energy. Under RIGES, by 2050, biomass could provide 17 per cent of electrical power and 38 per cent of direct fuel use. Altogether, "Renewable" could supply three-fifths of electrical power production and two-fifths of the direct fuel use, by 2050, at the same or lower cost to fossil fuel based-systems.

Biomass use has been finding better places as technical advances are making biomass a more promising energy source, and taking into account the environmental costs of power generation which all works in its favour. The most efficient power stations using biomass at the moment are "combined cycle" plants. These can turn more than 50 per cent of the energy in a gaseous fuel into electricity, compared with only 35 per cent in the best coal-fired plants. A French study has drawn attention to the employment implications: producing energy from biomass, it is reckoned,

Table 4.4

Summary of Regional and Global Projections for Ecologically Driven Scenario

Values are MTOE	Solar	Wind	Geo-thermal	Biomass (3) Trad.	Biomass (3) Modern	Ocean	Small Hydro	Large Hydro (1)	Renew. Total	% of Global (1)	"New" Total (2)	% of Global (1)
1990												
North America	3	1	6	38	19	0	4	127	198	2.3%	33	0.4%
Latin America	1	0	2	125	46	0	1	86	261	3.0%	50	0.6%
Western Europe	1	0	1	20	10	0	7	98	137	1.6%	19	0.2%
NIS & East Europe	2	0	0	30	10	0	2	53	97	1.1%	14	0.2%
MidEast/No. Africa	1	0	0	21	0	0	0	4	26	0.3%	1	0.0%
Sub-Saharan Africa	1	0	0	141	5	0	0	8	155	1.8%	6	0.1%
Pacific/China	1	0	3	351	23	0	4	70	452	5.2%	31	0.4%
Central/South Asia	2	0	0	204	8	0	0	20	234	2.7%	10	0.1%
TOTAL	12	1	12	930	121	0	18	466	1560	17.9%	164	1.9%
	0.8%	0.1%	0.8%	59.6%	7.8%	0.0%	1.2%	29.9%	100.0%			
2000												
North America	7	3	14	38	29	0	10	132	233	2.4%	62	0.6%
Latin America	2	0	4	140	86	0	2	150	384	4.0%	94	1.0%
Western Europe	2	2	5	19	14	0	15	101	158	1.6%	38	0.4%
NIS & East Europe	4	0	3	29	12	0	4	54	106	1.1%	23	0.2%
MidEast/No. Africa	2	0	0	25	4	0	0	8	39	0.4%	6	0.1%
Sub-Saharan Africa	1	0	0	190	15	0	0	14	221	2.3%	17	0.2%
Pacific/China	3	1	10	359	49	0	6	102	530	5.5%	69	0.7%
Central/South Asia	4	1	0	226	14	0	1	35	281	2.9%	20	0.2%
TOTAL	25	7	36	1026	223	1	38	596	1951	20.1%	329	3.4%
	1.3%	0.4%	1.8%	52.6%	11.4%	0.0%	1.9%	30.6%	100.0%			
2010												
North America	30	24	27	37	46	1	10	138	313	2.9%	138	1.3%
Latin America	8	0	6	144	163	4	3	215	543	5.1%	185	1.7%
Western Europe	11	10	9	17	24	4	17	104	196	1.8%	75	0.7%
NIS & East Europe	12	2	7	26	22	1	6	55	130	1.2%	49	0.5%
MidEast/No. Africa	7	0	0	27	7	0	0	11	53	0.5%	15	0.1%
Sub-Saharan Africa	6	0	1	222	29	2	0	20	280	2.6%	38	0.4%
Pacific/China	20	8	18	355	79	4	9	134	627	5.9%	138	1.3%
Central/South Asia	16	3	0	234	34	2	1	51	340	3.2%	56	0.5%
TOTAL	109	48	68	1062	404	18	47	727	2482	23.2%	694	6.5%
	4.4%	1.9%	2.8%	42.8%	16.3%	0.7%	1.9%	29.3%	100.0%			
2020												
North America	100	94	52	36	68	9	11	143	512	4.5%	333	2.9%
Latin America	39	5	12	144	286	11	4	279	780	6.9%	357	3.1%
Western Europe	31	37	15	15	34	5	19	107	262	2.3%	141	1.2%
NIS & East Europe	41	27	13	23	32	4	9	56	205	1.8%	125	1.1%
MidEast/No. Africa	29	6	1	27	11	0	0	15	89	0.8%	47	0.4%
Sub-Saharan Africa	30	4	2	239	48	6	0	26	355	3.1%	90	0.8%
Pacific/China	85	36	31	344	114	12	14	166	802	7.1%	292	2.6%
Central/South Asia	72	8	1	232	58	7	1	66	445	3.9%	147	1.3%
TOTAL	427	215	127	1060	651	54	59	857	3450	30.4%	1533	13.5%
	12.4%	6.2%	3.7%	30.7%	18.9%	1.6%	1.7%	24.8%	100.0%			

1) Based on ecology variant of new WEC M* scenario (with interpolation where necessary) – Global Total in 2020 is 11 340 MTOE
2) New equals Renewable Total less Traditional Biomass and Large Hydro
3) Biomass is divided into Traditional and Modern Categories

is three to six times as labour intensive as using fossil fuels.

D. Anderson in Report on Renewable Energy Technologies (1992) for the World Bank has emphasized three areas where policies that would begin to address the problem of global warming would also be beneficial to the developing countries. The first is the development of renewable energy sources wherein well-researched possibilities in the biomass and other sources of renewable energy exist. He has strongly recommended that the biomass technologies along with solar technologies have an immense promise, and may very well offer some economic and commercial surprise. In another study of the World Bank on "Renewable Energy Technology—A Review of Status and Cost" (1993) emphasis has been placed on three sources, namely, photoelectric, solar thermal and biomass for the production of electricity and liquid fuels.

Thomas B. Johannson et al in their book, *Renewables for Fuels and Electricity* (1992) published by Island Press have projected that the liquid and gaseous fuels derived from renewable energy sources for transport and other applications would be marketed much like oil and natural gas are today. Biomass grown in a sustainable manner would play a major role in both developing and industrialized countries. Biomass would be converted into electricity and liquid and gaseous fuels using modern conversion equipment that is far more energy efficient than the biomass conversion equipment now being used.

Robert H. Williams and Eric D. Larson in their paper "Advanced Gasification Based Biomass Power Generation" published in *Renewables for Fuels and Electricity* (1992) published by Island Press have reflected that a promising strategy for modernizing bioenergy is the production of electricity or the co-generation of electricity and heat using gasified biomass with advanced conversion technologies.

As regards the biomass production, productivity values of plantation have been reported to be up to 40 tonnes per hectare for high intensity plantation. However, continuous production facilities at 10-15 tonnes per hectare per year may be quite satisfactory.

The high yields achieved by intensive agriculture presently require energy intensive and mechanical methods of production. For certain crops under intensive management the energy required for cultivation and processing may exceed the energy content of the food produced. However, the energy output to input ratio of woody and fibrous energy crops is very favourable (in excess of ten times and about six times for ethanol from sugarcane). In general, cereal crops give a positive energy return even under intensive management, e.g. maize contains 3.5 times energy in the harvested grain than it requires to cultivate and process. Some crops presently required more energy to produce than they return in the food produced, e.g. energy output to input ratios are : apples 0.9, lettuce 0.2, tomatoes 0.6 and cabbage 0.8. However, these calculations do not account for the energy content of the associated residues which are significant. These figures are based on the recommendations of Pimentels, as quoted by J. Wood and D.Hall in 'Bioenergy for Development' - FAO Environment and Energy Paper 13 (1994).

The biomass provides an excellent opportunity to make farms energy exporters and net carbon sequesterers (sequester is defined as the net remover of carbon dioxide from the atmosphere and subsequent storage in organic matter).

Potential Global Productivities of Biomass

Total cropland increases have been at the expense of forests and woodland, with arable land estimated to have occupied 860 mha in 1982 and 1,477 mha in 1989. At the same time forested land has decreased from 5,200 mha, to 4,087 mha (31.2 per cent of the world's land area).

The production of agricultural surplus in Europe and North America has attracted large subsidies for lowering the production. Wright L.L, et al, in 'Opportunities to Mitigate Carbon Dioxide Build up Using Short-Rotation Woody Crops' in *Forests and Global Warming,* Vol 1 of American Forestry Association (1992) estimate that payment for the modern farm programmes in the USA is costing "one and a half times net farm income, whilst global agricultural subsidies were estimated to be about US $ 260 billion in 1990. If present policies continue as usual this will climb to US $ 300 billion by 2000".

It is estimated that in the United States about 30 mha have been removed from crop production by 1988. In the European Community (12 countries) surplus agricultural land resulting from rising yields and changing agricultural subsidies may reach 15-20 mha by the year 2000, and at least 40 mha in the following centuries if the productivities continue to increase. These croplands are already being removed from the intensive farming under set-aside schemes of the EC and "Cropland Reduction Scheme" of the USA. According to D. Hall this land represents a significant opportunity to initiate biomass energy production schemes, especially if coupled with the use of agricultural and forestry residues.

Estimates of degraded and abandoned land generally lie between 700 and 1,000 mha which is equivalent to about half the world's present arable land. The extent of this available land has led the scientists to highlight its potential for use in mitigating the Green House effect by managing it to become a carbon sink. Wastelands are regarded as having a good potential for storing carbon in trees due to the relatively low levels of carbon in their soils and vegetation.

According to D. Hall, at good productivities (6t /carbon/ha/year equivalent to about 12 oven dry t/ha/year biomass) the reforestation of all this land could theoretically remove about 5 GtC from the atmosphere per year over the next 40 years. Present emissions of carbon dioxide into the atmosphere are around 7-8 GtC/year of which 3-4 GtC appears as the atmospheric build-up in the levels of carbon dioxide. Sequestration programmes could, therefore, be regarded only as a temporary measure, buying time until other sustainable forms of energy or permanent carbon dioxide removal systems can be developed. It would require large quantities of inputs in the form of management, fertilizers, pesticides and labour in order to raise the productivity significantly.

Majority of degraded soils are salt affected. High salt levels reduce the levels of nitrogen which is available to the plant, but nitrogrn fixing species are often tolerant to saline soils, and may achieve acceptable productivities on such land. The potential for use of saline land is considerable, for example, salt tolerant plants can attain 3 to 7 tC/ha/year with saline water irrigation.

Role of Energy Plantation in Carbon Sequestration

Sustainably grown biomass is nearly carbon dioxide neutral, depending on production and conversion methods. If biofuel production is to be regarded as a a carbon sink, the total amount of carbon stored per hectare under the bioenergy crop must be geater than the level of stored C in the vegetation previously on that land . Also where biofuels are used as substitutes for fossil fuels, the biofuel can be regarded as a sink in terms of the avoided carbon dioxide emmision that would have been derived from fossil fuels.

The most clear benefit in carbon sequestration terms would be if the plantation was established in a desert with an existing standing stock of almost of 0 tC/ha. Both the above and below-ground levels of carbon would be raised significantly compared to the existing level. However, if the standing stock of the previous vegetation was greater than that of the new plantation then a net reduction in the standing C levels would result. If the plantation biomass were to be used as a fossil fuel substitute (or long-lived product), a net reduction to atmospheric carbon dioxide emissions would occur after a given time period. The higher the levels of vegetation, prior to the plantation, the longer it would take for fuel substitution benefits to reduce the amount of carbon dioxide initially released. This results from the relatively large amounts of carbon dioxide which are released during the initial clearance through harvesting and transport losses.

Marland G. and Marland S. (1992) in *Should We Store Carbon in Trees'* have explored the use of plantations for a fossil fuel substitution plantation-based model (1992). They conclude that the three most important factors in assessing whether biomass energy plantations are effective carbon sinks are as follows:

(i) The carbon inventory of the natural vegetation;
(ii) The productivity of the plantation; and
(iii) The time perspective adopted.

The most important points for the optimization of fossil fuel substitution are as follows:

(i) Increasing the energy output:input ratio, i.e. raising the ability of biomass derived fuels to substitute for fossil fuels; and
(ii) Raising the level of carbon held in the biotic pool in order to absorb some of the carbon already emitted by fossil fuel use, i.e. increasing the terrestrial carbon sink.

Biofuels provide an opportunity to avoid fossil-fuel- derived carbon dioxide emissions. However, the present intensive agricultural methods can be used only for the production of liquid fuels, eg, diesel from rapeseed, if the associated residue production is used for energy.Thus the usc of carbon-flows and energy output:input ratio to monitor net carbon dioxide production must be integral to the strategies for energy provision.

Bioenergy systems emit less sulphur dioxide (since they are naturally low in sulphur) and nitrous oxides than equivalent fossil-fuel-derived energy. Sulphur dioxide emissions are implicated in acid rain and increased nutrient depletion from soils .On the other hand sulphate aerosols, derived from sulphur dioxide emissions into the atmosphere have been shown to play a role as cloud condensation nuclei, and are thus postulated to moderate the global warming effect. Hence, increased use of bioenergy would theoretically reduce the levels of sulphur dioxide in the atmosphere thus removing this negative feedback effect, according to IPCC, 1992.

Increases in the levels of all major green house gases result from land clearing (devegetation), and thus well managed biofuels programmes which lead to revegetation will result in the decrease of gases. Indian Agricultural Research Institute established that paddy production in India is not the major contributor to the green house effect in the developing countries.

The secret of success of the bioenergy plantation lies in proper management of plantation from conceptualization to implementation.

TABLE 4.5 : Population Growth and Primary Energy Requirement in Selected Countries

	Total Population		Population Increase % P.A.		Primary Energy Requirement				Growth Rate	
	1990	2020	1990	2020	1990	2020	1990	2020	1990	2020
					Total (Gtoe)		Per Capita (Toe)			
Latin America	448.1	716.3	2.1	1.6	0.58	1.45	1.30	2.03	0.9	1.5
Sub Saharan Africa	501.6	1195.3	3.1	2.9	0.27	0.69	0.53	0.58	0.5	0.3
Pacific	1806.9	2428.4	1.5	1.0	1.85	3.53	1.02	1.46	2.0	1.2
South Asia	1146.0	1937.9	2.3	1.8	0.45	1.02	0.39	0.53	2.9	1.0

Projection of Fuelwood Use in the Globe — in Selected Regions

Decrease in the primary energy growth rate in the Pacific was due to an improvement in efficiency and partly due to higher population; and increase in South Asia was due to population increase.

TABLE 4.6 : Fuelwood Consumption (in Mtoe.)

	1990	2020	Change (1990-2020)	% Change
Latin America	125	179	54	43
Sub-Saharan Africa	141	299	158	112
Pacific*	351	414	63	18
South Asia	204	291	87	43

* Including centrally planned economies.

WEC Resource Handbook (1992) indicates that fuelwood plays an important and significant role in meeting the needs of developing countries.

Quantity of wood used directly as fuelwood and in the generation of the energy was 1.4 billion tonnes in 1990. This was just over half of all the wood consumed and corresponds to 0.5 billion toe. In addition, residue from the manufacture of wood products was recovered for the production of energy amounting to further 200 mtoe, making a total of 0.70 btoe. This is approximately 5 per cent of world energy consumption (as per FAO conversion rates). Consumption of wood for energy in the developing countries is 1.1 billion tonnes or 0.4 billion toe. In these countries, consumption of energy from fuelwood ranges from 0.1 to 0.5 toe per capita.

Worldwide some 2 billion people depend on fuelwood as their principal source of domestic energy. (This number also coincides with the number of persons deficient in some micro-nutrients in the developing countries.) The number is expected to increase by 50 per cent to 3 billion by 2000 AD and over 4 billion by 2020 AD.

It has been estimated that Asia and Africa would be largely dependent on fuelwood. In the case of Sub Saharan Africa, in some areas, dependence on the fuelwood is up to the extent of 80 per cent of total energy consumption. The more desirable end use forms of energy, such as electricity are still well beyond the economic means of the majority of its communities. Furthermore, considerations for commercial energy supply are far outweighed by the pressing needs for more basic services such as water, health, education and even "food security". Hence, wood will continue to be the dominant energy resource in spite of environmental, social and economic problems.

In case of South Asia, the large coal resources are likely to play an important role but hydro, biomass—mainly woody biomass—and biogas are seen as important sources of energy for the domestic sector. A balance of use of fuelwood, agrowaste and cowdung will continue to provide an energy mix in the rural areas in South Asia.

In the case of the Pacific region, which also includes China, Malaysia and the Papua New Guinea, fuelwood will still provide substantial quantities of energy. In China, some 80 per cent of the people living in rural areas live primarily on biomass energy forms, or on simple burning of coal, with unacceptable local environmental effects.

In the case of Latin America and the Carribean, fuelwood consumption is important, but its use is declining. It will thus be observed that fuelwood, mainly in the domestic sector, has an important role over quite a substantial geographic area which is also far more densely populated and poverty ridden.

Conclusion

Renewable energy systems in general, and bioenergy systems, in particular have an important role to play in the future global energy supply. Their contribution will depend on the development of proper management skills and adaptive technologies to optimize efficiencies in production and conversion through cost effective man and machine interaction.

5

GLOBAL BIOENERGY PERSPECTIVE

Energy and environment issues have dominated the global developmental scene, though the inter-linkages have been realized only recently. Growing concern for environment has made the whole U.N. system to move in the direction of ecologically sound developmental projects only. At the 14th Congress of the World Energy Conference held in Montreal, Canada, the Prime Minister of Canada, Hon'ble Mr. Brian Mulroney, in his inaugural address said, "We believe that environmental sensitivity and economic growth fuelled by energy go hand in hand. We no longer have the luxury of trying to have one without the other. But, we also believe that we are to have it both ways, as we must, we will all have to change our attitudes—and change them profoundly."

Therefore, the basic energy issues that confront the humanity are:

(i) Does the world have adequate economic energy resources to support the socio-economic needs of both the developing and industrialized countries?

(ii) Are energy resources in the form needed/available at reasonable prices to all countries wanting them?

(iii) Will R&D produce new energy supply options and more efficient use of energy?

(iv) When is such development likely to be applied on a significant scale? and

(v) Can energy be produced, transported and used in an environmentally acceptable manner?

Sixth Convention of BESI, Madurai, January 29-31, 1990

Oil to Dominate up to 2050

The global opinion indicates that oil will continue to be a dominant energy resource throughout the first half of the 21st century. The supply and utilization of oil will have the greatest social and economic impact, will undoubtedly be at the centre of geopolitical deliberations and will be a major factor in all environmental considerations. New technologies for finding and producing oil and gas are continuing to evolve. New technologies for enhancing recovery of oil in-situ are continuing to evolve and be applied at prices which may be acceptable soon or in a few years. The experts also have the opinion that in a market driven economy, the prices of energy sources will increase to provide for the exploration, development and production of future supplies.

Demand for Electricity in Developing World to Increase— But Who Will Fund Power Plants ?

Environmental considerations have once again supported nuclear power for electricity. While total energy use in the industrialized countries is growing slowly, the use of electricity has increased steadily due to its convenience and efficiency for a variety of tasks. The increased use of electricity in many of the smaller and poorer developing countries represents a particular problem due to the high cost of importing the necessary generating equipment as well as the generally high costs of the equipment utilizing electricity—most of which must be imported as well. In contrast to the development of some other forms of energy, the use of electricity does not directly produce foreign exchange to repay investments. This issue is not well understood despite numerous studies and the institutional assistance for the developing world is inadequate. The experts have the opinion that nuclear power will be increasingly recognized as an essential means of producing economic and safe electric power. This view has received significant impetus from the concern about the potential global warming because of the negligible emission of the carbon dioxide from the overall nuclear cycle.

The subject most discussed and related to the energy at the moment is—Environment (generally including health and safety as well); and the environmental issue directing the most attention was potential for global warming due to the accumulation of Green House gases, notably carbon dioxide. The importance of Green House gases having an impact on oceans and vegetation is now widely discussed.

The developing countries require relatively large amounts of energy per unit of economic progress, indigenous or imported, to improve their economies and living standards. While they are generally willing to minimize, to the extent feasible, the short-term and local environmental impacts of the energy production and use they are unable to forgo the use of available energy sources and still retain the possibility of economic and social development. It is completely unrealistic to expect this.

Electric power availability in the developing world over the next decade will require one trillion dollars on the global scale. The World Bank and other multi-national and bilateral sources, all put together, can raise only about 200 billion dollars. This reasserts the question as to where the remaining 800 billion dollars will come from?

Energy Demand

According to the WEC Study, world's population is expected to rise to 6,147 m in 2000 AD, and 7,829 m in 2020 AD. Major increases will be in the South where the population will grow to 4,765 m in 2000 AD, and 6,336 m in 2020 AD. Thus, the proportion which stands at about 75 per cent of the total global population of the South will gradually go up to 85 per cent of the population in the South. Average annual growth will be 3.2 per cent in the year 2000 AD and 2.8 per cent in 2020 AD. International energy price for oil in terms of 1987 US$ is expected to be $ 25 to 30 in 2000 and $ 40 to 50 in 2020.

Corresponding energy consumption between 1985 and 2020 would rise by about 50 to 75 per cent depending on the scenario, compared with a growth of 125 per cent during the period 1960 to 1985.

Energy savings compared with trends prior to 1973 amounted to 1 Gtoe or 14 per cent of world consumption in 1985. Savings could reach 6 to 10 Gtoe by 2020. Indeed, even by 2000, savings could reach around 3 Gtoe through demand management and would thus constitute the largest single energy source.

World's per capita consumption, which remained stable at around 1.5 toe during the period 1973 to 1985, would rise only slightly to 1.7 toe by 2020 for medium level growth projection. Energy intensity would continue the declining trend seen in the rates during 1973 to 1985. With the fall of 12.15 per cent per annum the demand in the industrialized North would increase gradually. In the South, on the other hand, the energy intensity situation, which showed no progress between 1973 and 1975, would improve considerably with an annual average of one per cent.

Third World demand would, therefore, increase from 2.3 Gtoe in 1985 to 4.9/5.8 Gtoe in 2020. As a result of demographic pressure, and relatively sustained economic development, the countries of the South would slowly catch up on the industrialized North in the share of world demand. Compared with 24 per cent in 1960 and 30 per cent in 1973, the South's proportion would rise to 35 per cent in 2000 and 43 per cent in 2020.

Renewable Energy Supplies

It would appear that sharply different trends would affect the various primary energy sources supplying to meet the world needs.

Non-commercial energy sources use, as the transfer to "modern" sources continues to lag, is set to increase in volume, in the South: 750 Mtoe in 1985 to 900/1,000 Mtoe in 2000; and 930/1,150 Mtoe by 2020. In percentage terms, NCEs would still be providing 15 to 25 per cent of the Third World energy needs in 2020 as against 33 per cent in 1985. The consequences of intensive use in agriculture, society and the environment are all too readily imagined.

Renewable energy sources including hydropower, new and non-commercial sources would enjoy only limited growth in the medium term on a global basis. Use of non-commercial energy sources is expected to increase in the developing world.

Non-commercial sources of energy like biogas, biomass, etc, abundantly used in the Third World are going to play a significant role in the South. The situation in Sub-Saharan Africa and Asia, the most disadvantaged region among the developing countries, continues to show alarming features. Consumption levels remain low and would be dependent for a long time on non-conventional energy sources (Table 4.1). These, in turn, would be increasingly in short supply and their intensive use would threaten the equilibrium of fragile eco-systems.

Availability of Firewood

The degree of dependence on firewood in the developing countries reaches up to 90 per cent in the case of the poorer countries. With the economic growth there is an increased use of petroleum and other fuels as well as a shift away from firewood. This leads to a decline in its proportionate needs of firewood in the national energy total, and ultimately, an absolute reduction in the quantities used. Fuel supply in the rural areas is usually the responsibility of women who collect what they need from their own farmlands and nearby woodlands. Firewood is often supplemented by the use of other biomass fuels such as dung and agricultural residues. Most of the urban firewood supplies on the other hand are commercially available and are obtained from natural forests or woodland sources. Firewood from these sources is extremely cheap because there are no establishment or maintenance costs. Charcoal is almost exclusively an urban fuel, but there is no obvious pattern in the distribution of its use around the developing world.

Wood is also used extremely widely as a fuel for small-scale rural industries such as tea and coffee processing, lime-making, brick-burning and other food processes and artisanal activities. It is also used on a small and medium scale for process steam and electricity generation. In a number of industrialized countries, the firewood related industries meet 30 to 50 per cent of their energy needs from residues.

Since most of the domestic use of firewood and other biomass fuels is outside commercial channels, it is extremely difficult to obtain reliable data on the amounts consumed. Sample surveys show that the quantities used by the domestic consumers vary widely and depend heavily on local availability. Overall world consumption of firewood for energy exceeds 1.2 billion tonnes or 0.45 billion tonnes of oil equivalent. Various new technologies were tried and all old technologies reused. In Japan, attempts to revive the charcoal gasifier technology for vehicles did not work due to high costs. In Finland, however, gasifiers which rely on wood or peat for fuel have been developed. There are seven district heating stations in operation with capacities in the range of 1.5 mw to 6 mw. A fluidized bed gasifier with a maximum output of 35 mw and using woodwool as feedstock was brought into operation in 1983 to supply fuel to lime kilns.

Growing and harnessing fuelwood for energy purposes are labour-intensive activities and the cost of fuelwood also depends heavily on local labour rates. In the industrial world, wood and charcoal are generally not competitive with commercial fuels. The exception is when wood is available as a waste product as happens in the industries.

TABLE 5.1: Forest Area and Fuelwood (including Charcoal) Production in 1987

Region/ Country	Total Forest Area million ha	Productive Forest Area million ha	Fuelwood (inc. Charcoal) Production ('000) tonnes
Africa	689	335	282687
Latin America	943	701	193631
North America	490	410	85696
Asia	351	231	389574
Australasia	50	40	2092
CPEs	944	706	229357
Western Europe	108	95	33571
Middle East	5	3	2345
World Total	3581	2521	1218953 (0.45 Btoe)
India	59	45	164384

In the developing world, the majority of the firewood used for fuel comes from natural woodland for which there are no planting and maintenance costs. In consequence, commercially viable fuelwood plantations are rare. Attempts have been made to create urban peripheral plantations to meet urban fuel demands in a number of countries but the wood produced has not been able to compete with that from natural woodlands or, often, with conventional fuels such as coal and petroleum.

It was formerly believed that firewood cutting was a major cause of the deforestation taking place throughout the developing world. It is now accepted that the expansion of agriculture is the main reason for the loss of forests. Even if the cutting of wood for fuel were to stop completely, the rate at which deforestation is taking place would not be greatly reduced.

Availability of Biomass Other Than Wood

The potential of biomass as an energy source should be seen in perspective from the fact that the annual photosynthetic storage of energy in the form of biomass is ten-fold the use of energy from all sources throughout the world. Biomass fuels represent the dominant energy source for the developing countries, providing more than one-third of their total energy demand, and the main domestic energy source for 2.5 billion people. In some countries, e.g. Tanzania, 90 per cent of their energy comes from biomass. In China, its 800 million rural population derived 84 per cent of their energy from biomass (equivalent to about 328 million tonnes coal) in 1981. Although firewood is a predominant energy source in many of the poorest nations, it is by no means always the case. In India, energy from burning dung and agricultural residues is equivalent to energy from firewood, resulting in a total biomass contribution of about 50 per cent to India's total energy supply. In China, the weight of available dung and agricultural residues is 2.2 times the weight of firewood. The use of biomass on a large scale is not confined to the developing nations. In the USA today, for example, biomass contributes nearly 4 per cent of the primary energy — this is greater than the contribution of the nuclear. It is anticipated that the biomass contribution will double by the year 2000.

Important advances have already been made which have contributed to a significant increases in (a) plant productivity through improved clones, agronomic techniques, intercropping practices, agroforestry, etc; and (b) biomass utilization through greater technological efficiency combined with low pollution.

Processes for the conversion of biomass to useful energy generally fall into two categories: (i) biological; and (ii) thermochemical.

The present utilization of biomass for energy generation is mainly through direct combustion, fermentation to alcohol, and anaerobic digestion. Though in many cases the stage of development limits the applicability, many of these processes are being continuously improved. Local circumstances determine which process for biomass utilization is most appropriate, depending on the resource, the conversion facility and end use requirements. Examples of biomass use for foreign exchange savings, for generation of employment and for providing economic incentives for management of natural resources are seen in Brazil, Zimbabwe and China.

Brazil, since the creation of the ethanol programme at the end of 1975, has produced over 85 x 10^9 litres of ethanol and over 3.5 million vehicles now run on ethanol produced from sugarcane; the rest, about 9 million vehicles, run on a blend of 20 to 22 per cent ethanol. The ethanol programme has been largely successful in meeting the technological objectives of increasing agricultural and industrial productivity in related sectors, increasing technological capability, reducing oil imports, and creating significant employment in the rural sector. Sugarcane productivity has increased from an average of 46 tonnes per hectare in 1977-78 to 60 tonnes in 1985-86. Fermentation efficiency improved from about 75 per cent in 1975 to 83 per cent in 1987. Recent eucalyptus plantations have shown, on average, an annual mean yield of 55 m^3 of wood per ha/year and in some cases over 100m^3 per ha/year against an average of 36m^3 per ha/year a decade ago.

6

BIOENERGY APPLICATION - A GLOBAL OVERVIEW

Bioenergy resources are contributing substantially meeting the energy needs of the developing countries, especially in the domestic sector in the rural areas. The 13th Congress of the World Energy Conference held at Cannes, France, during 5-10 October, 1986, also emphasized this aspect. Though the fuelwood was not discussed in detail its contribution in the overall energy supply in the developing countries was emphasized.

The Round Table on "Rural Energy in Developing Countries" was of the view that "the rural population in the developing countries (and therefore an important percentage of the world population) has only a limited access to traditional sources of commercial energy. They are more dependent on non-commercial sources like firewood. This situation may change in the long-run but that may take several decades". Awareness of the extreme importance of this problem began in New Delhi, three years ago, at the 12th Congress of the World Energy Conference in 1983, and was confirmed at Cannes, France.

Various discussions m7ade it clear that the energy used in the developing world is for the most part biomass-based energy and decisions are made by women, for which action to be taken lies not only in the technical field but in the social one too.

Energy Use Pattern in Developing Countries

In the developed Western economies (depending to some extent on their climatic conditions) over 90 per cent of the energy consumption derives from commercial fuels, i.e. oil, gas, solid fuels (coal and lignite) and water

---

Third Convention of BESI, Pune, January 28-30, 1986.

falls. This energy is extracted, processed and transported commercially and mechanically. In other words the user has the fuel or energy commodities delivered in processed form at the point of consumption. In many developing countries, e.g. Nepal, Malawi, Ethiopia, etc, over 90 per cent of energy consumption derives from traditional non-commercial fuels, i.e. fuelwood, agricultural waste, dung, etc. Such fuels normally have to be extracted, processed and transported by the consumer himself — or more normally — herself. In the developed Western economies the annual per capita commercial energy consumption amounts to between 2 tonnes and 8 toe. (tonnes of oil equivalent). In countries like Nepal the annual per capita commercial energy consumption amounts to 13 kilogram of oil equivalent. This is less than one-five-hundredth of what the average European consumes. If we include non-commercial fuels, then we find that the average Nepali will consume about 1.5 cubic metres of fuelwood, equivalent to about 400 kilograms of oil equivalent or one- twentieth of the average of Western economies.

Given the economic and socio-economic realities of energy production, it is not difficult to discern women's involvement in the energy scene in the developing countries. In the developed Western economies virtually all activities requiring the use of energy have been commercialized and mechanized, permitting an enormous per capita consumption of energy in relation to the situation in the developing countries. In the developed countries, however, the situation is reverse. In the first place, virtually none of the activities requiring the use of energy have been commercialized or mechanized. This means that the consumer (in 90 per cent of cases the women of the family) must also be producer, processor and transporter. However, the situation is even worse than that. In many developing countries there is a lack of diesel oil for pumping water because of foreign exchange shortages, and because of the general lack of commercial energy forms, the bulk of the work which might, in other circumstances, be mechanized, e.g. agricultural labour, cannot be mechanized. This creates a situation in the developing countries whereby the production of non-commercial fuels, e.g. fuelwood, requires an undue amount of human labour, walking up to 10 km a day, felling trees and transporting headloads of wood. Daily agricultural production also requires undue amounts of energy and labour inputs. Such a situation puts a heavy burden upon the women of society, particulary in societies where women already have an inferior social and cultural position.

World Energy Survey

The role of biomass resources including firewood was considered extremely important. The Survey of Energy Resources prepared for the World Energy Conference, 1986, made various recommendations identifying the resources available in each country in terms of proven and additional reserves as far as alternative sources are concerned. The Survey concerned itself mainly with the achievement to date and current projects. From the data provided it was possible to derive estimates for the world as a whole and to compare reserves between countries and regions.

The Survey is consistently being developed to take account of opposing terms in energy reserves. On the one hand, rising consumption and currently falling prices are bringing about progressive reduction in stocks and, on the other, research, new discoveries and an inexorable rise in prices in the long-run allow existing stocks to be revised upwards. Alternative or renewable energy sources have been included indicating that the statistics of thc field are still at an early stage. Their impact is expected to be gradual.

The fact that the current situation is not particularly urgent means that the difficulties of tomorrow tend to get ignored. Of course, such difficulties will be more serious if governments, energy producers and consumers start to obtain the positive energy policies that do not redress the balance of equilibrium of the world energy system.

This Survey obviously was hinting at the recent turn- about in energy policies by the industrialized world due to sharply declining oil prices. The experts were of the opinion that even if the prices are going down there is no reason for the energy consumers to be careless and start consuming the fossil fuels at a much faster rate. The policies of tapping various new and renewable resources of energy must be vigorously pursued as they were done in the past. However, the economies in the changed circumstances where the price of oil has crashed below $ 10 per barrel does not permit many of the R&D works to be pursued. A large number of industrialized countries have already backed out and the money is becoming scarce.

The energy situation in the developing world where biomass can play an important role must be looked into, especially from the point of view of the increasing population. It is projected that at present the developing world is having 75 per cent of the total world population whereas they arc consuming only about 25 per cent of the energy in the world. By the year 2020 the developing countries will be having 80 per cent of the total population and if the present trends are continued the energy availability will be on the lower side. It was, therefore, suggested that the energy availability for the developing countries must be substantially increased and roughly one-third of the total energy consumption for the year 2020 should be in the developing countries. This does not mean that there will be a sudden change of policies and the energy could be transferred from the developed world to the developing world, but it is basically a hint to the developed countries for the energy policies to be followed in the near future.

The Survey mainly concerned itself with conversion and utilization and not with the production of biomass resources.

Future Development: Growth Pattern and Energy Availability

FAO has projected the concept of Primate City based on growing urbanization and the challenge of rural migration. They have projected that all net additions to the world rural population and 84 per cent of the net additions to the world urban population between 1980 and 2000 would be in the developing world. During this period it is likely that nearly 500 million migrants will move from rural to urban areas in the developing countries resulting in raising the percentage of the urban population from 31 to 44 per cent of the total. Thus, two-thirds of the net additions to the population of developing countries will reside in urban areas by 2000. Most will be without land to grow their own food. Yet the high population growth rates in the developing countries also mean that the rural population will increase further. It is projected that by the year 2000 over 2,700 million people in the developing countries will be living in rural areas. It is also projected that 21 out of 25 of the world's largest cities will be in the developing countries by the year 2000, compared with 13 out of 25 in 1970.

Thus, the situation in the developing world is going to be much more complicated than that in the developed world. It is clear that the developing nations have to keep both their options open — on the one hand meeting energy requirements of the urban population basically on centralized systems like large capacity electricity plants and modern transportation systems; and, on the other, continue feeding their rural population from various traditional biomass-based small energy producing systems. The developed world does not have to face any such problem. It already has centralized systems, with well-connected grids for meeting most of the needs for electricity. The technology will not be easily available to the developing countries and products bought will be at an exorbitant price which the economy of most of the nations cannot afford.

Therefore, it was suggested that the developing countries must continue their options of utilization of indigenously existing resources in the most efficient fashion and develop them further so that the rural population is able to meet its requirements fully .

Options in Specific Areas

Discussions at the World Energy Congress 1986 led to three options for use of biomass :

(i) Biomass in the form of Firewood.
(ii) Biomass in the form of Ethanol Fuel.
(iii) Biomass in the form of Gas by Gassification or Anaerobic Digestion.

(i) Biomass in the Form of Fuelwood

Forests make up 4,100 million ha or 30 per cent of the world's land area. They are distributed between: the temperate zone, with 2,650 million ha, of which about 1,650 million ha are closed forests and 550 million ha of open woodland; and the arid zone, which contains some 300 million ha of forests, mainly open woodland. There are, in addition, about 1,000 million ha of land classified under uses other than forest with some shrubs or woody vegetation.

Some 11 million ha of forest are lost each year as they are cleared for agriculture, destroyed or degraded by shifting cultivation, fuelwood gathering, overgrazing and burning. The total tree planting of 1 million ha a year in tropical countries make up for only one-tenth of the area of forests annually cleared and destroyed. The uncontrolled destruction of forest in the humid tropics and arid zones is of major concern both for the future supply of forest products in these regions and for the maintenance of the forests' conservation functions. Solving this problem involves a complex series of issues, including improved technical understanding and the resolution of conflicting pressures for the use of land.

The annual production of roundwood from forests is around 3,000 million cum. Nearly one-fifth of this quantity provides raw material for mechanically processed wood products as materials for housing construction, furniture, packaging and communication, or for paper. About 80 per cent of the 1,300 million cum produced in the developed countries is used for such industrial purposes. However, the same proportion of the wood cut in the developing countries is used as fuel (1,377 million cum in 1983), with Africa depending for more than one-half of its total energy supply on fuelwood. Other forest products include fruits, gums, resins, fodder and meat of wild animals.

Firewood production is being increased by identifying quick growing tree species. Though only four or five varieties are propagated there can be a larger number of site specific varieties that would enhance the total firewood production. Every country was expected to identify such varieties. India has identified 32 varieties for different agro-climatic zones.

Fuelwood growing itself can be employment generating. Efforts must be made to link this programme with other rural employment or development programmes so that people have an economic attraction also. An example was cited of the Energy Plantation for Power Programme in the Philippines. Originally this programme envisaged employment for 100 persons, but now 400 persons are employed.

Similarly the demand for fuelwood in Ougadougou, the capital of Burkina Faso, has provided an additional source of income to small-scale farmers in the nearby accessible areas. Increased demand for fuelwood-based energy will enhance the economics of growing wood for fuel. However, adequate management of this source of income calls for the allocation of increasing resources for investment in planting and conservation to ensure future supplies.

Most of the rural applications require a low quantum of conversions whether for direct burning as in wood-burning stoves or small size gasifiers of 5 kw range. Participants were of the opinion that low-range gasifiers need a lot of research work. The "FAO Network on Small Size Gasifiers" was seen as an important step.

Use of agricultural residue and waste that can be converted into useful energy were also projected. Details of worldwide availability of these resources was also projected.

Although the use of fuelwood in the developing countries also expanded, its contribution to total energy has declined, particularly in Asia. Africa remains the region depending most heavily on fuelwood for its energy supply, but even in this region, the contribution of fuelwood has declined.

(ii) Biomass in the Form of Ethanol Fuels

Ethanol produced by fermentation of suitable substrates, such as elementary carbohydrates (derived from materials containing sugars, starch and cellulose), is a carrier of solar energy. Solar energy is first trapped in the chemical bond of energy of biomass by the action of photo-synthesis. The elementary carbohydrates can be directly converted by the enzymatic varieties of yeasts into the ethanol and carbon dioxide. Ordinary yeasts are not able to convert higher sugars such as those found in starch and cellulose containing materials into ethanol. These materials are hydrolysed to elementary sugars prior to fermentation into ethanol.

A major difficulty in analysing the effect of ethanol fuels is the lack of reliable and truly comprehensive statistics regarding the ethanol production, consumption and trade.

No major technical obstacle prevents the commercial utilization of ethanol fuels as is attested by the large-scale use achieved in many countries.

(iii) Biomass in the Form of Gas by Gasification and Anaerobic Digestion

Biomass resource utilization in the gaseous form is most important and common in the tropical developing countries. Gasification of woody biomass and anaerobic digestion of biomass, including residue, is equally important. Though gasification is comparatively a newer technology, recent developments have resulted in wider acceptance and application. Gasification units of lower sizes are more common. Its most important role possibly is in its capacity to replace diesel in diesel-electric sets. Upto 80 per cent replacement has been practically observed, though 55 per cent is normally achieved in the field.

Looking into the importance FAO has also set up a Network for Small Gasifiers. A large number of operation level nodal agencies in different countries are its members. This coordinated approach is likely to solve most of the problems of the developing countries. Users of small-scale gasifiers have been facing a number of problems that are expected to be solved through this Network. As for now the application of these gasifiers in rural areas will increase substantially.

The other route in the gaseous form is biogas. It is generated from biomass and is utilized throughout the world, but the state of technology varies according to the raw materials available and the emphasis in local economies. Currently there are millions of domestic digesters in use in different parts of the world. Emphasis is leading China from domestic services to community to industrial scale operation. In Latin America, the adoption of anaerobic digestion has been more selective, but has benefited from a wealth of agricultural wastes. Biogas generation is a mature technology in Europe, but implementation is critically affected by local economic constraints.

7

PERSPECTIVE PLANNING FOR BIOENERGY IN INDIA

Energy has been playing an important role in the national economy of the country. Different forms of energy like electric power, fossil fuel power and renewable sources of energy have been systematically utilized under different climatic, economic and social categories of the country. Indian economy is such that a major part of its population is still living in rural areas and are slowly making use of the latest in science and technology. As a matter of fact India has a fabric of mixed system where the jet airliner and bullock- cart co-exist.

Forestry in India

Forestry in India is a State subject and the State Governments administer forests. But, the Union Government exercises supervisory control. It has its own Department of Forestry under the Ministry of Environment and Forestry. The concept of managing forests in India is now growing although the natural forests have provided a wide green coverage to the landmass. Growth of forestry and the laws and acts covering it have been changing from time to time. Some of the significant aspects can be referred to as follows:

(i) *Scientific Forestry*

The practice of scientific forestry in India may be said to have begun when Sir Dietrich Brandis took over as Inspector General of Forests, India, in

---

Bio Energy Users' Network, First Membership Meeting, Bangkok, October 1-5, 1985.

1864. It was at his instance that the Government decided to adopt a Scientific Policy for Forests. On the advice of Dr. Voelcker, a German expert, the Government of India issued a resolution dated October 19, 1864, declaring a Forest Act Policy. In 1865, the Indian Forest Act was passed.

Under this Act the Indian forests were surveyed and classified as reserved and protected. In the reserved forests, the existing rights were settled, transferred or commuted, while in the protected forests they were recorded and regulated. To put it otherwise, in a reserved forest, everything was an offence that was not permitted while in the protected forest, nothing was an offence that was not prohibited.

The Forest Policy declared by the Government laid down the broad principles under which forests of different types were to be administered. Successful implementation of this excellent forest policy resulted in a steady development of the forests in the country. For nearly 60 years, forests were continuously conserved. There was some depletion of the forest capital during the First World War, but this did not affect the growth and accumulation of forest capital. But during the World War Second forests were heavily felled far beyond their sustained production.

(ii) *National Forest Policy*

When India became independent the Forest Policy of 1894 was reviewed in the light of later experience and a National Forest Policy was promulgated in 1952.

The National Forest Policy resolution proposed that the area under forests be steadily raised to 33.3 per cent of the total land area, the proportion to be aimed at being 60 per cent in the hilly areas and 20 per cent in the plains.

In the years that have elapsed since the National Forest Policy was declared, two important developments have taken place side by side that have a significant bearing on forestry. The rapid growth of population had increased the demand for a large variety of forest products, firewood for example, and at the same time had led to the occupation of more forest lands for diverse purposes. Between 1952 and 1973 about 34 lakh hectares of forest land were occupied for such purposes as cultivation, river valley projects, industrial estates, etc. In December, 1973, the National Development Council decided that no further deforestation is to be allowed for expansion of cultivation and that in order to compensate for the land already lost suitable wastelands should be afforested. The Forest (Conservation) Act of 1980 embodies the resolution of the National

Development Council. The Act ordains that notwithstanding anything contained in any other law for the time being in force in a State, no State Government or other authority shall make except with the prior approval of the Central Government any order directing:

(i) that any reserve forest or any portion thereof shall cease to be reserved; and
(ii) that any forest land or any portion thereof may be used for any non-forest purpose.

According to the then Union Minister for Agriculture, more than 4 million hectares of forest land had been deforested in the last 30 years. The new Act was expected to put a stop to similar encroachments in future.

(iii) *Forest Area*

India's total forest area is 75.06 million hectares which is 22.88 per cent of the total geographical area of the country. The per capita forest area works out to 0.11 hectare which is extremely low. India's forest area constitutes 1.85 per cent of the world forest area (4,057 million hectares) whereas the population forms 15.64 per cent of the world population (4,335 million).

(iv) *Forestry Resources*

India's total forest area is 74,786 million hectare and is normally not categorized as is done in the marketing economies of the Western world.

(v) *Categories of Forests*

Indian forests fall under five botanical types:

(a) *Evergreen Forests* are found where the rainfall is 200-300 cm, i.e. the Western Ghats and the sub-Himalayan regions in the east. Hard wood such as teak, ebony, rosewood and ironwood are found in this area. Bamboos of all varieties thrive well here.
(b) *Deciduous Forests* or monsoon forests are seen in areas where the rainfall ranges from 150 to 200 cm. These forests occupy the larger part of the Deccan Plateau. The trees are big and provide fine timber—teak, sal, padauk, sandalwood, redwood, ariun, etc.
(c) *Dry Forests* occur where rainfall is between 75-100 cm. The desert regions of Rajasthan and South Punjab are such places.

TABLE 7.1: Forestry Resources in India (1976-77).

(in '000 ha)

Sr.No.	Description	Area
1.	Geographical Area	328,779
2.	Forest Area	74,786
3.	Forest Area as a Percentage of Geographical Area	22.75%
4.	Net Area under Management by Working Plans	49,659
5.	Potentially Exploitable Area	15,995
6.	Net Area under Management (in use)	45,061
7.	Growing Stock ('000 M^3)	1,964,416
8	Annual Availability ('000 M^3)	30,067

TABLE 7.2: Estimated Sustained Yield of Wood (Recorded Outturn of Wood in m cu.m)

i.	Natural Forests - high forests - woodlands	1964.416	
ii.	Forest Plantations	1975-76	1976-77
	a. Timber Plantations	7,991,000	8,292
	b. Fuelwood Plantations (coniferous and non-coniferous)	19,284,000	20,311

TABLE 7.3: Skills Bank Information

1.	Estimated Growing Stock	:	1896 m cu.m.
2.	Exploration Rate	:	1.5%/year
3.	Outturn of Fuel and Timber Wood	:	32 m cu.m. (1979-80) 13.5 m cu.m. (Industrial) 18.5 m cu.m. (Fuelwood)
4.	Fuelwood Consumption	:	127 m cu.m. (FAO 1983)
5.	Plantation	:	2.63m ha (Eucalyptus, Teak, Sissoo, Bamboo, Casuarina, etc.)
6.	Average Annual Output	:	1.44 m cu.m. (1961-65) 5.62 m cu.m. (1981-85)

The plants here have thick and fleshy stems and leaves, and the forests mostly consist of thickets of shrubs with a few stunted trees.

(d) *Hill Forests* are found in South India, above 1,500 metres, and in the Himalayas, above 1,000 metres elevation. Trees are evergreen and include the very best timber like oak, picea, deodar, pine, chestnut, walnut, maple, elmash and birch; laurels, poplars and rhododendrons also grow profusely here.

(e) *Littoral Forests* are found in the estuaries of great rivers like the Indus, the Ganges and the Mahanadi. The trees here provide excellent firewood and constitute a rich source of fuel.

Forest Development Programmes

Forest development programmes stress what are called *production forestry* and *social forestry*.

Production Forestry: It concentrates on raising man-made forests to meet the long-term requirements of forest products in general and materials for wood-based industries in particular.

The Government of India has taken the lead in establishing State Forest Development Corporations, with central participation. These corporations are to be financially assisted by term lending institutions in India and wherever possible by the World Bank and its associates. The corporations are expected to take up massive operations.

Social Forestry: It aims at augmenting fuelwood resources and minor wood for farmers by new plantations of fuelwood trees and small trees suitable for farm needs.

Minor Forest Products

All forest produce other than wood are considered minor forest produce. These include bamboos and canes, grasses of various kinds, leaves of different trees, medicinal plants, gum, resin and lac.

Bamboo: The most important and plentiful plant among minor forest products is the ubiquitous bamboo. It is seen in many varieties from the giant bamboo of the Assam forests to the dwarfs of the dry Deccan forests. The two main types are male bamboo and thorny bamboo.

Thorny bamboo flowers gregariously in about 30 to 40 years and dies. Once the entire bamboo dies after flowering, it takes some years for fresh bamboo clumps to appear.

The flowering of the male bamboo is regarded as a portent of famine all over India. Actually, in north-eastern India the flowering is often followed by famines. The connection between the two is obscure. It is presumed that rodents feed on the profusion of bamboo seeds on the ground (both in public forests and in private lands) and multiply. The greatly increased rodent population eats up the cereal grains. This causes a food shortage, which may lead to famine, unless prompt remedial measures are adopted.

Forest Leaves: Leaves of certain forests are of commercial value. Leaves of the climber Mahul can be used for making cups and plates and are commercially disposed off as such. The Tendu leaf is by and large the most important forest leaf. This is used as the wrapper for beedis and is in great demand all over India. All States where Tendu plant grows count upon them as a major source of income. Tendu trees are found in plenty in Madhya Pradesh and Orissa, which are the major producers of this product.

Gums: These are used as mucilage for sizing in confectionery or as medicines. The most important gum- yielding trees are Babul, Bijasal, Kulu, Salat and Dhaora. Of these the gum called Kareya from Kulu trees is used for mixing with ice-creams to thicken them. It is exported and brings in valuable foreign exchange.

Resins: They differ from gums in that they are soluble in alcohol. The most important resin in India is that of Chir. The tree is tapped in Punjab, Himachal Pradesh and U.P. The pines yield turpentine, which is now being extracted on a commercial scale.

Katha: Katha is an Stringent substance obtained from a species of plant called *Acacia catechu* locally known as *khair. Katha* is valued as a medicine for digestive problems. It is also used as an ingredient in paint. Cutch which is a by-product of *katha* is an excellent tanning material.

About 3,000 tonnes of *katha* are produced annually in India. The leading States are Uttar Pradesh, Madhya Pradesh and Bihar.

Medicinal Plants

There are over 20,000 species of medicinal plants of proven therapeutic value in the Indian forests. They are collected and used as drugs and medicines under the Indian medical system.

Energy Scene in India

The present per capita commercial energy consumption in India is only about one-tenth of the global average, reflecting the country's low level of income. While India, in common with many developing countries, consumes energy in a variety of forms, ranging from electricity obtained from nuclear fuels to domestic energy obtained from agricultural wastes and animal dung, the pattern of commercial energy consumption is characterized by a high degree of oil dependence, the share of oil products being close to 50 per cent.

The economic growth has to be sustained at planned levels, for which adequate energy supplies need to be arranged. This obviously leads to new sources of energy being developed at a rapid pace. Identification of research and development in emerging energy technologies, exploitation of renewable sources of energy, like energy forestry, biogas, solar energy, etc., and energy conservation are some of the areas that will require attention in due course of time. It has also been observed that different sources of energy can meet demands in a site specific application, like in the rural areas. Cow- dung, firewood, etc, will be the sources that would be found much more appropriate for household applications. Though reliable data is not available rough estimates indicate that between 40 and 45 per cent of the total energy consumed is still accounted from non-commercial fuels.

India's endowment of fossil energy resources is by no means large. There are also obvious limits to the extent to which savings in energy consumption could be achieved through conservation. It is apparent, therefore, that if the country's growing demand of energy is to be met, new and renewable sources of energy will have to be developed. The new energy technologies are particularly suited for the introduction of decentralized or small-scale energy supply systems which fit in India's rural agricultural economy. Remote communities which cannot be reached by electricity grido excepting at a prohibitive cost or which do not have easy access to conventional commercial fuels, could, through renewable energy technologies, be provided with energy sources for domestic use as well as for rural industry and agriculture. In general, these technologies have the additional merit of having a less adverse impact on the environment than the conventional fossil fuel systems. Even though the immediate contribution of the new energy technologies to energy supply may be limited and many of them are not economically competitive, at the moment, there is no doubt that they will play a predominant role in the future.

TABLE 7.4 : Commercial Energy Consumption Pattern
(in percentage)

Sector	Share of Each Sector	Coal	Oil	Electricity
Domestic	15.00	1.50	10.68	2.82
Agriculture	10.00	-	6.18	3.82
Industry	41.00	18.25	3.24	19.51
Transport	34.00	4.52	28.53	0.93
	100.00	24.27	48.63	27.10

Whereas the conventional energy technologies are characterized by long gestation periods, it takes even longer to bring about significant changes in the pattern of energy consumption. Appropriate policy and investment decisions have therefore, to be taken without loss of time, if the country is to meet the challenge, which the energy problem will pose in the coming decades.

Plans for Renewable Energy Programme in India

With this backdrop, research and development, fully supported by the extension programme in the field of new and renewable energy, has been taken up in India. The broad approach for the development of new and renewable sources of energy is as follows:

(1) To implement on a large scale programmes such as those of energy forestry and biogas where technology development has already reached a stage which permits field application;
(2) To carry out field testing and demonstration, on a countrywide basis, of technologies which have the potential to become commercially viable in the next five to seven years; and
(3) To intensify research and development of other technologies where the potential is likely to be available over a longer time horizon.

In the emerging energy situation it is extremely necessary to sustain the supplies of non-commercial fuels and to improve the efficiency with which they are utilized. Firewood, it is well known, is the most important among the traditional fuels, accounting for almost two-thirds of the total energy contribution from non-commercial sources. It is becoming increasingly scarce and many parts of the country have already ceased to be non-commercial in the strict sense. The ecological damage caused by an extensive denudation of forests for fuel and other requirements is too well known to need any elaboration here. There is, no doubt, the problem of pressure on land and the competition for land use for agriculture, industry and housing. Recent experience, however, shows that even in the thickly populated areas of the country sufficient land is available along the roadsides, railway lines and canals and in the degraded forests, and community land for developing woodlots to provide fuel to the local communities. Energy forestry is being recorded as an important component of our energy strategy, as the rural communities continue to depend heavily on firewood for several decades to come.

Development of Bioenergy Resources

The bio-energy applications receiving a thrust are the following:

Biomass: The total amount of biomass produced, per annum, in India is about 1,249 mt. Assuming the energy equivalent of biomass as 4,400 kilo calories per kgm, the total latest energy comes to about 525 giga-kilo calories. This latent availability can be increased by improved photosynthesis and also by developing species which can thrive on saline soils and in unfavorable environment.

About 40 per cent of this potential can be used as a bioenergy resource. This will be equivalent to 350 mt of wood or 175 mt of kerosene, in terms of energy.

Biogas: Biogas plants have become an established feature of the energy scene in India with around 4.55 lakh plants already installed. Of these 1.8 lakh were installed during 1984-85. With an average size of 4 cum these plants deliver about 18.20 lakh cum of biogas per day on a plant utilization factor of 300 days a year.

A common family size of 4 cum capacity plant produces about 1,200 cum of gas and 20 tonnes of manure per annum. As 1 cum of biogas is equivalent of 3.47 kg of firewood or 0.62 litre of kerosene, the quantity of firewood saved works out to 4.16 tonnes, or kerosene saved as 744 litres per annum per family. Thus, if 3 lakh biogas plants are installed every year, 1 million tonne of firewood (after providing 20 per cent allowance) can be saved, every year. This equals the increase in demand due to increasing population. The major plant nutrient contents in the manure from plants would be of the order of 200 kg of nitrogen, 100 kg of phosphorous and 100 kg of potash per year, which would be adequate for manuring between 2 and 3 ha of cropland.

The biogas programme has now really become a common man's programme and the number of complaints is gradually coming down.

Smokeless Chulha: Smokeless wood burning chulha is an important device. Over 7 lakh improved *chulhas* have already been installed and over 5,000 villages made smokeless. Practically observed information reveals that the quantity of fuelwood burnt in *chulhas* can be reduced by almost 30 per cent by scientific designing. However, in one of the experiments carried out by us in village Achheja, it was observed that the quantity of fuelwood saved was 40 per cent.

If we assume that in practical application we are able to save 40 to 45 per cent of the firewood by using efficient chulhas, deforestation due to firewood burning can be arrested.

Energy Plantation

The main objective of the Energy Forestry/Plantation Programme in India is to produce fuelwood rather than industrial wood. Therefore, selection of species for different agroclimatic conditions has assumed greater significance. The specific objectives for the selection of species are:

(i) larger biomass field;
(ii) high calorific value;
(iii) fast growth;
(iv) good coppicing (regeneration after harvest);
(v) high adaptability
(vi) ease of establishment;
(vii) little aftercare;
(viii) identification of plants for problem regions; and
(ix) species with nitrogen fixing ability.

The whole strategy for development of plantation is covered by diversity in agro-climatic conditions, like soil type, rainfall, etc. Problems, therefore, become location specific. The cost of implementation will vary from place to place and fuelwood production from species to species.

With a view to carrying out investigations on screening, evaluation and agronomic trials of fast-growing firewood species under different agro-climatic conditions, three Biomass Research Centres have been set up by the Department of Non-Conventional Energy Sources, Government of India. These are:

(i) National Botanical Research Institute, Lucknow (for Saline/ Sodic Soils);
(ii) Madurai Kamraj University, Madurai (for Arid and Semi-arid Regions); and
(iii) Garhwal University, Srinagar (for Hilly and Mountainous Regions).

The main objectives of these studies include selection of fast-growing species, with a short rotation cycle; and establishment of plantation trials for assessing the yield potentials under a variety of agro-silvicultural practices.

A number of demonstration centres have been set up under the Energy Plantation Demonstration Programme. These centres are created with the objective of awareness creation of the role of fast-growing fuelwood species in meeting the energy needs of particular areas. A number of projects have been started under this programme in different parts of the country. These include Yeotmal (Maharashtra); Yellamanchalli, Visakhapatnam (A.P.); Aligarh (U.P.); Bhavanagar, Khandia, Distt. Vadodra (Gujarat); Pipalaya Phandekhan (M.P.); Kilakuchi (T.N.); Chandnagar (A.P.); Gopannapalli Rambadi, Thanda (A.P.), etc. The implementation of the programme is spread over 100 ha of sub-standard soils and is being implemented in close cooperation with the State Forest Corporation, Forest Departments, Nodal Agencies and Research Institutions.

This programme is to be extended to over 1.5 million hectares of sub-standard soils like saline alkali soils, water- logged soils, hilly soils, ravine lands, shifting sand dunes, laterite and lateritic soils, black soils and coastal alluvium spread in different regions—moist tropical, sub-tropical, dry tropical and temperate regions of the country.

The choice of firewood species and the technique of growing will be both site specific in character. However, these will depend, to a large extent, on a thorough understanding of soil characteristics. Thus, the role of soil scientists in this programme is equally important.

Future Potential Bioenergy Resources

A substantial amount of energy depending on crop-land needs for food production could be produced from biomass sources in India. This energy could come from numerous types of biomass, including wood, grass and legume herbage, grain and sugar crops, crop residues, animal manure, food- processing wastes, oil-bearing plants, kelp from ocean farms, and many other materials. But the overwhelming majority of this energy would come from woody or lignocellulosic materials such as firewood from commercial forests, various types of herbage, especially grasses and legumes, from existing pasture land and hayland and crop residues. Consequently, if India wishes to exploit the total potential of biomass energy in the next 20 years or so the process for converting firewood, grasses and crop residues to usable energy should be emphasized, and ways of harvesting and collecting these materials must be promoted that will avoid severe environmental damage. Because of the difficulty of collecting large quantities of these materials in a single place, considerable emphasis will have to be placed on process designs that may be applied in small-to-medium scale facilities.

The major processes for converting solid biomass fuels to more usable energy forms are direct combustion, airblown gasification, and alcohol fuels synthesis. The principal concerns about harvesting these materials are:

(i) that wood from existing forests be collected in a way that maintains the long-term productivity of forest-land for all of its uses, and increases do not hinder the production of timber suitable for lumber and paper pulp; and
(ii) that sufficient crop residues be left in place to protect soil from excessive erosion.

Power Generation from Energy Plantations

Utilization of sub-standard or non-agricultural land for planting fast-growing species of trees having a rotation cycle of 4-5 years only is an important feature of an energy plantation programme for power generation. It has been estimated that about 1,000 hectares can provide 3 mw of power, besides providing fuelwood or charcoal which can support the energy needs of a population of 125-150 families. According to Mr. M. Dayal, Secretary, Department of Non- Conventional Energy Sources, Government of India, even if one-third of the estimated 80 million hectares of barren and wasteland can be recovered by such a programme, a generation capacity of about 48,000 mw can be created, which is considerably more than the entire installed power capacity in the country today from all thermal, hydro and nuclear sources. Moreover, power from energy plantations could be available from a number of decentralized installations, thus minimizing transmission and distribution costs. This would also provide fuelwood and charcoal to meet the cooking needs of the rural poor, provide green biomass cover in arid zones and raise rural income. The investment cost for such projects could be only about Rs.12,000 per kw.

According to a calculation, the cost of electricity produced in such a programme would be about 87 paise per kwh, assuming the capital cost of local generation at Rs.12,000 per kw, fuelwood is priced at Rs.50 per quintal, O & M costs 10 paise, an annual capacity factor of 0.60 and a rate of return of 0.12. If the fuelwood is priced at Rs. 25 per quintal (since it is available next to the power unit) the cost of electricity comes down to 62 per kwh. This can be favourably compared to a cost of 96 paise per kwh for electricity from a central power station, to a village having a load of 20 kw, and at a distance of 10 km from the grid; these costs increase to Rs. 1.11 per kwh at a distance of 20 km, and Rs. 1.26 per kwh at a distance of 30 km from the grid. Thus, power from energy plantations is a very promising option for substantial power needs.

Firewood Tree Species for Different Climatic Zones

The Fuelwood Study Committee Report of 1982 has identified 34 crop species which can be grown in different agro-climatic regions of India. Of these 30 species have been identified by the botanists, foresters, plant explorers and others as multi-purpose plants.

Different species suggested here for the utilization to the fullest extent are grouped under four eco-geographical zones:

(i) humid lowland tropics and subtropics;
(ii) semi-arid tropics and subtropics;
(iii) tropical montane areas; and
(iv) coastal region.

These species are as follows:

I. *Humid Lowland Tropics and Sub-tropics*

The following species would be suitable for cultivation in humid Western Himalayan regions; humid Bengal and Assam region; humid Eastern Himalayan region and Bay of Bengal Islands; and subhumid Sutlej-Ganga Alluvial Plains.

1. *Acacia catechu Wild ("Cutch tree", "Khair")*
A medium sized gregarious, thorny, deciduous tree. The wood weighs 54 to 76 lbs per cubic foot and calorific value of hard wood is 5,244 (9439

BTU). The calorific value of soft wood is 5,142 cal. (9,256 BTU).

2. *Anogeisus latifolia Wall ("Axlewood")*
A large deciduous tree with a smooth, whitish-grey bark which has shallow, irregular depressions. The calorific value of soft wood (CVS) is 4,986 (8,976 BTU) and that of hardwood (CVH) is 4,909 (8,837 BTU). The weight of wood is 62-65 lbs per cubic foot.

3. *Artocarpus heterophyllus Lamk. ("Jack-fruit")*
A large evergreen tree, with alternative, ovate-oblong to ellipticoblong, glabrous, acuminate, entire leaves. The wood weighs 43-45 lbs per cubic foot. Its MCS is 4.13 per cent; MCH is 7.07 per cent; ACS is 2.24 per cent; ACH is 2.36 per cent; CVS is 5,267 cal. (9,482 BTU), CVH is 5,369 cal. (9665 BTU).

4. *Azadirchta indica A. Juss. ("Neem" or "Margosa tree").*
A large beautiful evergreen tree. The wood weighs about 50-52 lbs per cubic foot. It is a quick growing tree and coppices well.

5. *Butea monosperma Taub ("Pulas kino tree", "Dhak", "Phulla", "flame of the forest")*
A moderate-sized deciduous tree. The wood weighs about 35 lbs per cubic foot. Its MCS is 13.45 per cent; ACS is 3.12 per cent; CVS is 4,925 cal. (8,865 BTU).

6. *Cassia fistula Linn. ("Indian Laburnum", "Kinjal", "Krinjal")*
A moderate-sized deciduous tree. The wood weighs about 52-73 lbs per cubic foot. Its MCS is 12.77 per cent, MCH is 12.18 per cent, ACS is 1.35 per cent, ACH is 0.61 per cent; CVS is 5,164 cal. (9,296 BTU), CVH is 5,164 cal. (9,296 BTU).

7. *Dalbergia sissoo Roxb. ("Sisoo", "Shisham")*
A large deciduous tree. The wood weighs 45-60 lbs. per cubic foot. Its MCS is 10.56 per cent, MCH is 9.93 per cent, ACS is 2.69 per cent, ACH is 2.40 per cent; CVS is 4,908 cal. (9,835 BTU), CVH is 5,181 cal. (9,326 BTU).

8. *Eucalyptus globulus Labill. ("Tasmanian blue gum", "Kurpoora maram")*
A lofty tree, with fibrous deciduous grey outer bark. The wood weighs about 46 to 64 lbs per cubic foot. Its MCS is 12.34 per cent; ACS is 0.25

per cent; CVS is 4,962 cal. (8,932 BTU).

9. *Gmelina arborea Roxb. ("Gumhar", "Khammara")*
A moderate-sized or large deciduous tree. The wood weighs 30 to 40 lbs per cubic foot. Its calorific value is 4,763 cal. (8,574 BTU) for soft wood).

10. *Hardiwickia binata Roxb ("Anjun")*
11. *Kyida calycina Roxb. ("Pola", "Barange")*
12. *Leucaena leucocephaic Wit. ("Horse tamarind", "vilaiti haval")*
13. *Madhuca indica J.F. Gmel. ("Mahua tree", or "Mathua") Bassia latifolia Roxb.*
14. *Morus alba Linn. ("White mulberry", "tut")*
15. *Phyllanthus emblica Linn. ("Emblica")*
16. *Pongamia pinnata Pierre. ("Pongain', "Korans", "Indian beech")*
17. *Syzygium cumini Skeels. ("Jamun", "black-plum tree")*
18. *Tamarindus indica Linn. ("Tamarind tree", "imli")*
19. *Trema orientalis Bl. ("Charcoal tree", "Indian nettlet tree", "jivan")*

II. **Semi-Arid Tropics and Subtropics**

20. *Acacia nilotica subsp, indica Brenan* (Syn. A. arabia Willd ("Babul", "Kikar").
21. *Ailanthus excelsa Roxb.* ("Tree of heaven", "arua", "maharukh").
22. *Cassia auriculata Linn.* ("Tanner's bark", "tarwar", "Aiwal").
23. *Prossopis spicigera Linn.* ("Jhand", "Khar").
24. *Zizyphus mauritiana Lamk.* ("Jujube", "ber").

III. **Tropical Montane Areas**

25. *Bosewellia serrara Roxb. ex Colebr.* ("Indian frankencense", *luban, sallai*).
26. *Dodonaea viscosa Linn. (Jamaica switch, sorrel, vilaiti mehnidi, jalpai).*
27. *Malltous philippensis. (Kamala dye tree, kamela, kamal, kambal).*
28. *Moringa oleifera Lamk.* (Syn. *M. Pterygosperma Gaertnn)* (Drumstick tree, horse-radish tree, *sujuna).*
29. *Callicarpa arborea Roxb. (Ghiwala, dera, shiwali).*

IV. ***Coastal Region***

30. *Anacardium occidentale Linn* (Cashew nut tree, *Kaju*)

Barriers, Constraints and Problem

Virtually all forms of bioenergy suitable for energy can have non-energy uses as well, and bioenergy production will compete with other uses for the same land base. If care is taken to integrate energy with non-energy objectives, the estimated energy potential from wood and plant herbage probably can be obtained without severe competition from non-energy uses of these materials. The Government of India is seriously considering utilization of total barren and waste lands. Most of these lands are saline or alkaline in nature and are not fit for agricultural production.

Whereas production of firewood by use of quick growing trees is a safe and preferable route, the one based on production of petroleum from plants and shrubs is not ideally suited for India.

If wood energy harvests are a part of a comprehensive silviculture programme, they can actually increase growth of timber suitable for lumber and paper pulp.

Competition between Energy and Non-Energy Uses of Biomass

Competition between energy and non-energy uses of biomass as well as other uncertainties can affect reliability of fuel supplies. Wood and plant herbage supplies may be diverted from non-energy uses, that is, at times, have a greater economic value. Adverse weather conditions also can interrupt harvesting or reduce total biomass productivity per acre. In addition, in areas where biomass fuels have been put to use, imbalances can arise between quantity produced and consumption needs. Moreover, many of these factors could cause the problem of high transportation cost of bioenergy. Hence bioenergy systems that use oil or natural gas as back-up fuel look particularly attractive.

Because biomass fuels tend to be voluminous and of low fuel value per pound, their transportation costs, related to other fuels, are high. These costs and the dispersed nature of the resources may limit the size of bioenergy facilities to those requiring less than 1,000 tonnes of biomass fuel per day (roughly equivalent to the input of 60 mw electric generating plant). Therefore, market penetration would be aided by the development of reliable automatic, and inexpensive, smaller conversion systems — especially mass produced gasifiers or high efficiency incinerators — so that small industrial, residential and commercial users who are familiar with oil, gas or electricity can switch to biomass without having to learn new skills or major changes in their operation.

Environmental Benefits

Major *potential environmental benefits* of biomass energy development are:

- (i) constructive use of wastes that could otherwise cause pollution;
- (ii) opportunity to improve forest productivity relieving logging pressures on some environmentally fragile lands; and
- (iii) replacement of environmentally more harmful energy sources, especially coal.

Biomass energy development may include:

- (i) substantial increases in soil erosion and in sedimentation of rivers and lakes and subsequent damage to land and water resources;
- (ii) adverse changes in or loss of important ecosystems;
- (iii) degradation of aesthetics and recreational values;
- (iv) local air and water pollution problems; and
- (v) occupational hazards.

These damages, although not inevitable, appear likely to occur for a number of reasons:

- (i) First, some of the less intensive agricultural and forestry operations from which biomass supply mechanism could be derived already cause serious pollution problems;
- (ii) Second, biomass feedstock suppliers as well as conversion facilities may be hard to regulate because the choice of appropriate control and management techniques is very site specific, making it difficult to develop effective and enforceable guidelines for environmental protection. There are also likely to be a great multitude of small sources, thereby creating a significant monitoring and enforcement problem;
- (iii) Third, the existing economic and regulatory incentives for biomass suppliers and users to protect environment are weak; and
- (iv) Finally, some of the currently most popular biomass alternatives —alcohol from grains and woodstoves for residential heating—

have a high potential for environmental damage. The major dangers from grain alcohols are the erosion and ecosystem displacement that would be caused by expanding crop acre able to increase production, while large increases in woodstove use may lead to serious public health problems from particulate air pollution.

It has also been suggested that long-term losses in forest or crop productivity will result from declines in soil organic matter associated with residue removal and high-intensity (short rotation, whole-tree harvesting) forest management. However, the degree of these impacts is somewhat speculative at this time.

Alternative biomass feedstocks have sharply different possibilities for environmental damage.

Social Benefits

Biomass energy development is likely to be more labour intensive than increased production of conventional fuels (coal, natural gas, oil) in the short-term. Thus, increases in employment due to bioenergy will occur in resource harvesting (agriculture and forestry); manufacture, distribution, and servicing of conversion equipment (gasifiers, boilers, stills, anaerobic digesters, wood stoves); and construction and operation of large-scale conversion facilities (generating plants, alcohol fuel plants).

Due to biomass fuel transportation costs, most of these employment increases will arise at small, dispersed rural sites near the resource base. If bioenergy development becomes a major contributor to U.S. energy supplies, the new jobs could alleviate unemployment and under-employment among rural residents — especially in agricultural and forested areas — shift the rural age distribution to a younger population, and help to revitalize rural areas. Agricultural areas, in particular, will benefit from the degree of liquid fuels self-sufficiency afforded by on-farm distillation as well as the overall energy contribution from anaerobic digestion. Moreover, if commercial alcohol fuels' production is managed properly, the nation as a whole would benefit from reduced dependence on imported oil. On the other hand, to the extent that bioenergy is subsidized, it will attract investment and jobs at the expense of other sectors.

However, increased bioenergy production also could result in increased rates of accidental injuries and deaths in energy related occupations. Consequently, safer biomass harvesting and conversion methods need to be developed and implemented. Occupations associated with bioenergy (logging, forestry, agriculture) generally have higher occupational injury rates than do jobs in conventional fossil fuel sectors (coal mining, oil and gas extraction).

8

ENERGY PLANTATION FOR POWER SMALL CAN BE BEAUTIFUL

The oil price escalation in 1975 once again focussed attention of the developing world on the need for new and renewable sources of energy. Many exploration works and researches taken up in this field would never have been undertaken if crude oil were to sell at $ 3 per barrel, the price in 1973. This situation has thrown a challenge to the scientists involved in biomass research.

One of the major areas that has received specific attention is energy plantation, especially for power generation. Whereas it will help to substantially augment energy resource, it will also help in maintaining the eco-balance by providing additional forest cover. Forests, because of their protective, regulative and productive functions at the primary eco-system level, are the super constituents of the biosphere. The important protective functions are: (i) protection of soil from radiation, precipitation and wind; (ii) conservation of soil and atmospheric humidity; (iii) regulation of oxygen and carbon dioxide in the atmosphere and mineral elements in the soil; (iv) absorption, storage and release of water; (v) provision of shelter for diverse flora and fauna; and (vi) transformation of radiant thermal energy.

Many of these functions are of direct utility to man. These include efficient storage of energy in usable form in the biomass; self-regulative and regenerative process of wood, bark, fruit; and protection of a wide variety of other items which are of great use to people. Judicious use of these resources coupled with balanced rural development has been

---

Workshop on Biomass Availability and Assessment Techniques organized by INCORE, Visakhapatnam, June 29-July 1, 1985.

recognized as the core of national development strategies and programmes.

"Indian forests" like "Indian cows" are venerated but neglected. Whereas the cattle population has increased, the forest area has gradually diminished. We have even less than one-fifth of the area of the country under forest cover when the requirement as per the National Forest Policy is one-third.

Until such time as the satellite imagery generated data to assess natural resources, the total land area under forest cover was more or less a guess work, especially its contraction and a shift could not be easily assessed on a real; time basis. Satellite imageries in the 70s and 80s show denudation of forest over a decade. The situation is really alarming when we look at the figure of 0.1 hectare of forest available per capita as against the world average of 1 hectare. In spite of the focus on social forestry and aforestation programmes, the situation has not improved.

This situation is aggravated by the effort to meet the firewood requirement for domestic and other industrial needs. The present need (in 1985) of firewood is 138 m tonnes against the availability of only 49 m tonnes. It is expected that by the end of 1985 the total availability of firewood might increase to 73 m tonnes, still leaving a gap of 65 m tonnes to be taken care of. Added to it is the fact that in the coming years the quantity of firewood required would gradually be on the higher side, at least at a rate of 1m tonnes per year. Thus by the end of the next decade as we march into the 21st century, India will need over 160 m tonnes of firewood. In areas of acute scarcity of firewood, agricultural waste, animal dung, twigs and leaves, etc., are burnt resulting in a drop in soil fertility. Use of cow-dung as fuel in an inefficient way is estimated to result in a loss of additional 45 m tonnes of agricultural crops.

The scientific imagery process has given us a tool or an aide to realize the gravity of situation created by higher consumption of firewood in our country. Though the age-old method of physical mapping has provided certain records in the revenue departments and the forest departments, these are never updated periodically to project a realistic picture to the planners to inable them scientifically plan future progress of the country. New techniques and methods evolving at INCORE are a boon at a time when a massive programme of utilization of waste and barren lands (in 1985) is being undertaken at the rate of 5m hectares per year or 25 m hectares in the Seventh Five Year Plan. This is one of the most ambitious and massive programmes — a challenge which is being looked upon by the whole world with extreme curiosity.

The existing infrastructure of the forest department was able to ensure reforestation of 0.45 m hectares in the year 1984-85 and at present is geared up to plant 1 m hectares per year. This 1 m hectare mark is significant as India is losing forest cover at the rate of 1 m hectare per year for firewood alone. In case we are able to achieve the target of 5 m hectares per year even by the end of the Seventh Five Year Plan it should be possible for us to achieve even higher targets and cover 120 m hectares of waste and barren lands by the turn of the century.

Energy Plantation for Power

Looking in this context the programme of energy plantation for power, EPP, is extremely significant and important. Recent indicators are that major demonstration programmes to instil confidence in the people are being undertaken by the Department of Non-Conventional Energy Sources to provide better forest cover on the one hand and energy for rural needs and employment opportunities on the other. The fast-growing species which can be used in a rotation of 3 to 4 years are being selected for this purpose. The genetic improvements in the species will certainly enable to produce a larger amount of biomass per unit area. The EPP programme can be extremely effective in the rural context for the simple reason that plantations in an area of 1,000 hectares of barren land can generate 3mw of electricity besides providing fuelwood and charcoal to support and fulfil the energy needs of about 125 families. In addition almost 100 persons will be directly employed in the plantation and power generation programme. Thus 1,000 hectare programme provides a package which gives employment to one person per family, from over 100 families, and also provides them with sufficient charcoal/fuelwood for their domestic needs. This is rural development based on decentralized energy availability.

This projection does not mean that plantation on a lower side will not be feasible. The only thing is that plantation of this size has proved economically viable in certain South Asian countries. We, in India, can always make "small look beautiful" and going in for plantations of about 50 hectares of land generating some 150 to 200 kw of electricity. On an average 100 kw of electricity can substantially meet the basic needs of almost 10,000 population of the rural areas scattered in 6 or 7 conglomerates, or generate enough commercial activity along with meeting energy needs.

Since this programme is directly linked to the well- being of rural people and can, in due course, become an important part in the total rural development scheme, it would be appropriate to consider the new avenues that local availability of electricity can open by way of village industries, better irrigation facilities, pure drinking water supply, educational facilities and primary health medical care, in addition to communication systems either one-way, like TV and radio, or two-way link through microwave telephony systems.

A Case Study of Bisrakh Block

Since the new concept in rural development requires planning at district and block levels and then adding to make national rural development plan, the Indian Association for the Advancement of Science undertook a programme to assess the barren land availability and to study the possibility of growing biomass in Bisrakh Block having village Achheja in District Ghaziabad of Uttar Pradesh. This study was conducted by Shri A. K. Tyagi, Project Officer, NEDA, and Dr. V.P.S Sorayan of the Centre for RD&AT, IIT, Delhi. The blockwise break-up of barren and wasteland in District Ghaziabad is given in Table 8.1.

However, confining our discussion only to Block Bisrakh, the barren land availability in this Block is 747 ha. If all this was available at one place then we could theoretically generate 2.3 mw of electricity. This land is scattered over the whole block and further sub-divided into three categories as follows:

(i) Forest land;
(ii) Public land; and
(iii) Private land.

TABLE 8.1 : Blockwise Break-up of Usar Land in District Ghaziabad, Uttar Pradesh

	Name of Block	*Usar Land available +(in Hectares)	Possibility of Use for Energy Plantation
1.	Garh	1,713	Yes
2.	Simbhaoli	120	Yes
3.	Hapur	283	Yes
4.	Bhojpur	1,739	Yes
5.	Dhaulana	2,205	Yes
6.	Muradnagar	309	Yes
7.	Rajapur	752	Yes
8.	Loni	642	Yes
9.	Bisrakh	747	Yes
10.	Dadri	1,987	Yes
	Total	10,497	

* Saline land.

Since the EPP programme is conducted by the Department of Non-Conventional Energy Sources, the engagement into the land owned by the Forest Department is at the moment ruled out. Of the remaining, part of the land is under private ownership and what is available as public land "is roughly about 300 hectares" which is just sufficient to generate a total of 1 mw of electricity. Various locations where these lands are available are normally in pockets of 10 to 25 hectares. Thus a strategy is being finalized to club together the nearabout barren lands which will at least total to about 30 to 40 ha. so that 100 kw of electricity generation could be sustained.

Some of the figures for this block as such can be extremely interesting. Bisrakh can on the whole generate almost about 1 mw of electricity by promoting the EPP programme. The investment cost for such projects could be only Rs.12,000 per kw and the cost of electricity produced would have worked out to about 62 paise per kwh in a decentralized fashion with local load centres. Since land is in small holding the cost will be slightly higher, may be about 70 to 75 paise. This compares favourably with a cost of 96 paise per kwh electricity from central power station to a village having a load of 20 kw and a distance of 10 km from the grid. This cost increases to Rs. 1.11 kwh at a distance of 20 km from the grid.

Conclusion

This is just a small example of how biomass can be assessed and explored effectively. Estimates show that our barren and wastelands can generate up to 48,000 mw of electricity.

Thus, this type of assessment at block level in a micro fashion and then their totalling up at the district level and the State level can really prove a boon for decentralized electricity generation systems for the rural areas as well as for meeting domestic energy needs and creating rural employment opportunities for the rural masses.

One should not, however, get an impression that this type of projection once undertaken for a block can be replicated for every other block in the country. At the same time we should also realize that the data now current as regards barren and wastelands may become outdated in a few years' time and unless a total system of "satellite imagery, physical mapping and verification is undertaken", the schemes may go haywire.

9

ENERGY PLANTATION AND DESERT DEVELOPMENT

Arid Lands Can Generate 30,000 mw

Almost 10 years back the world's nations agreed to a sweeping Plan of Action to half the process of destruction called desertification. UNEP completed a two-year assessment in 1984 which revealed that the global threat posed by desertification, far from diminishing, has actually increased in severity. Also about 35 per cent of the world's land surface was considered at risk and the livelihood of 850 million people who live there was directly threatened. Three quarters of the 45 million sq km that make up the world's drylands, which include the sub-humid tropics, is already affected, between half and a quarter severely so.

This means that a total of 3,475 million hectares of the world's range, rainfed cropland and irrigated land — an area approximately the size of North and South America combined — is affected by desertification. Currently each year some 21 million hectares is reduced to a state of near or complete uselessness though only 6 million hectares come under desert classification. Projections for the year 2000 indicate that a loss on this scale will continue if nations fail to step up remedial action.

In case of desertification the main cause is not drought as many still believe (drought tends to exacerbate the problem) but human over exploitation of lands through over cultivation, overgrazing, poor irrigation practices and deforestation. And in turn, the underlying causes of these lie in bad management, rural neglect and political and economic forces resulting mainly from the world's inequitable financial arrangements and

---

Fourth Convention of BESI, Udaipur, September 4-5, 1987.

terms of trade.

Desertification results not only in the loss of nations' productive resource base but also in the loss of valuable genetic resources, increase in atmospheric pest, disruption of natural water recycling process, loss of markets, etc.

Though this problem has been realized by the rulers in the world, they are not finding sufficient financial resources to fight this menace. The absurdly inadequate level of contributions to the special account set up in 1979 to finance such programmes is an illustration of the low priority nations attach to tackling the problem. The Plan of Action to Combat Desertification (PACD) was agreed to by 94 nations in 1977. The special machinery, the General Assembly set up to mobilize funds to tackle desertification, had raised, in its first six years of existence, only $ 26 million, 25 per cent of the minimum target figure. These sums should be weighed against the $ 4.5 billion needed annually over a period of 20 years to stop desertification.

Compared to this if an energy plantation programme is taken up the world will be a much happier place to live in and especially the arid lands will provide a better living environment to the rural population. One estimate of UNEP in 1984 reflects that in the year 2030 global rural population will probably be 9 per cent higher than now, while that of all developing countries will be 17 per cent higher, although in Africa, in most of its arid countries, the increase will be nearer to 200 per cent.

Desertification — Indian Scene

The problem of desertification was mentioned in the First Five Year Plan as "Indian desert of Rajasthan has been spreading outwards in a great convex arc at the rate of about half a mile per year for the last 50 years." In recognition of this problem, the National Institute of Science (now Indian National Science Academy) organized a Symposium on the Rajputana Desert in the year 1952. On its recommendations Desert Afforestation Station was established. It was later named Central Arid Zone Research Institute (CAZRI).

The Government of India and State Government in the arid and semi-arid zones realized the alarming situation and reacted promptly.

The arid and semi-arid zone in India is spread over eight States, but 90 per cent of the hot desert is located in north-west India, out of which 61 per cent is located in the State of Rajasthan (see Table 9.1). It is by far the most densely populated desert of the world, and the manifestations of environmental deterioration are particularly evident in this region. Also it is possibly the most studied desert in the world.

TABLE 9.1 : Areas of the Arid Zone in India by State

	Area Under the Arid Zone (sq km)	Percentage of the Total Arid Zone in India
Rajasthan	196,150	61.0
Gujarat	62,180	19.6
Punjab	14,510	5.0
Haryana	12,840	4.0
Maharashtra	1,290	0.4
Karnataka	8,570	3.0
Andhra Pradesh	21,550	7.0
	317,090	100.0
Jammu & Kashmir (cold arid zone)	30,300	

Source: A. Krishnan: *Desertification and Its Control.*

Regeneration Need

Besides climatic fluctuations, erratic rains and extremes in diurnal temperatures, the principal problem of the Indian desert is the escalation of human and livestock population. Most of the other problems are in reality the consequences of these two major factors causing environmental degradation and desertification.

The human population in arid zones has escalated from 3.56 million in 1961 to 13.94 million in 1981. The livestock population has also registered an increase from 9.4 million in 1961 to 19.8 million in 1977. The devegetation and degradation of the arid lands due to the presence of large human and livestock populations has necessitated the restoration and maintenance of vegetative cover both to stabilize and protect soils and to provide fuelwood to desert people and fodder to the livestock. Keeping these objectives in view, CAZRI has developed and standardized a number of cost-effective technologies for revegetating the Indian desert as mentioned below:

(i) Sand Dune Stabilization
(ii) Afforestation
(iii) Shelter Belt Establishment

(iv) Arid Horticulture
(v) Range Management
(vi) Soil and Moisture Conservation Measures
(vii) Re-seeding
(viii) Introduction of Legumes
(ix) Silvo-Pastoral Approach

Desert Revegetation Programmes

In 1970-71 the Government of India sponsored a Rural Works Programme for organizing labour-intensive and production-oriented works in the drought-prone districts of the country, in addition to the normal developmental efforts implemented in these districts under the State Development Plans.

The scope of the programme was subsequently enlarged to form an integrated area development scheme to seek a permanent solution of the problems of drought in these districts. Accordingly, the emphasis shifted from labour- oriented to problem-oriented schemes. This programme was designated as the Drought Prone Area Programme (DPAP) in 1972-73.

The Government of India launched another programme in 1977-78 designated as the Desert Development Programme with the objective of controlling desertification, restoration of ecological balance and creation of conditions which would improve the economic level of people in these areas. These objectives are sought to be achieved through activities, such as afforestation with special emphasis on shelter belt plantation, grassland development and sand-dune stabilization, measures aimed at conservation of surface water and recharge of groundwater acquifer and efficient water management through appropriate structures and practices suited to the area and development of agriculture, horticulture and livestock resources suited to the agroclimatic condition of the area.

Coverage of the programme has been extended to 131 blocks in 21 districts of 5 States from the beginning of the Seventh Plan, i.e. 1985-86. It now covers about 3.62 lakh sq km with a population of about 15 million, constituting about 42 per cent of the area and 16 per cent of the population of the States in which the programme is operative. This is being implemented as a fully funded Central Sector Scheme.

The physical achievements during 1985-86 and 1986-87 (up to December, 1986) are as given in Table 9.2.

TABLE 9.2: Achievement of Desert Revegetation

	Sector	Achievements (ha) 1985-86	1986-87 (up to Dec. 86)
i.	Forestry plantation including roadside plantation, fuel-wood and fodder and pasture development	18,656	7,144
ii.	Irrigation potential created	971	871
iii.	Soil and Water conservation	5,115	2,638

Future Strategy

The strategy for development of desert areas has to take into account special problems experienced in such areas. As the shifting sand dunes pose hazards to the adjoining cultivable land, sand dune stabilization assumes priority in development activity. Equally important is shelter belt plantation which provides protection to the field crops, prevents/arrests soil erosion and restoration of ecological balance.

Desert areas have considerable cattle wealth but, due to scarcity of fodder and drinking water, the cattle has to migrate to other areas. Since availability of water is a crucial factor in taking up the above-mentioned development programmes, judicious exploitation of ground and surface water consistent with ecological requirements and accompanied by measures for conservation of rainwater and facilitating recharge possibility of groundwater acquifers are of paramount importance. This calls for the most efficient and economical water management practices. Serious efforts will be made through extension methods to promote agronomic practices, cropping pattern, species of trees, livestock development pattern which demand least water and moisture. In close cooperation with the concerned agencies integration will be sought with other programmes like NREP and RLEGP, Rural Drinking Water Supply Scheme, Anti-Poverty Programmes of IRDP and other State and Central schemes.

There is no infrastructure for technological research in cold deserts. This has hampered the pace of development in these areas where problems are vastly different from those in the hot deserts.

Energy Plantation for Greening the Deserts

A number of steps that can reduce desertification have been mentioned above. However, one route is through bioenergy plantation. This may be most appropriate as these plants need very little moisture in soil and can withstand adverse climatic conditions. In other words the possibility of survival of suitable energy plantation species is better than any other horticultural crop or tree.

Energy plantation cannot cover the whole desert but can significantly contribute to greening of the desert. The trees so grown will become a local source of energy for gasification or direct burning to produce electricity and charcoal for domestic energy needs.

It has been estimated that plantation on 1,000 ha can support generation of 3 mw electricity for 8 hours a day from the fourth year of plantation and employ over 200 persons.

Even if one-third of the total arid lands can be put under energy plantations, over 10 million hectares can yield over 137 million tonnes of firewood, provide employment to over 10 lakh people and sustain electricity generation cap°city of over 30,000 mw. (See details in Table 9.3).

Moreover, the source of energy so available will help in setting up rural industries and improving the quality of life. Thus, the employment opportunity shall be much higher.

Some of the species found suitable for arid zones are listed separately, in Table 9.4.

TABLE 9.3: Yield of Wood, Cost of Energy Plantation and Potential for Electricity Generation on Arid Lands

Name of State	1/3rd of the Total Arid Lands (in ha)	Yield of Fuelwood in Tonnes	Cost of Energy Plantation (in m Rs)	Potential for Electricity Generation (in mw)
Rajasthan	6,531,795	91,445,130	65,318	19,595
Gujarat	2,070,594	28,988,316	20,706	6,211
Punjab	483,183	6,764,562	4,832	1,449
Haryana	41,840	5,765,760	4,184	1,255
Maharashtra	4,296	59,584	430	120
Karnataka	285,381	3,995,334	2,854	856
Andhra Pradesh	717,615	10,046,610	7,176	2,152
	10,134,704	137,065,296	98,324	31,638

Notes

(i) Yield has been taken as 14 tonnes/ha (Estimate by the Gujarat Energy Development Agency, GEDA).

(ii) Cost of energy plantation has been taken as Rs.10,000 ha. (The norm set by the Department of Non-Conventional Energy Sources for Demonstration Projects)

(iii) 1,000 hectare to result in 3 mw is taken as potential electricity generation (the norm observed in Energy Plantation Programme in the Philippines).

TABLE 9.4: Plant Species Suitable for Energy Plantation in Arid and Semi-Arid Zones

Sr.No.	Genetic Name	Popular Name
1.	*Acacia arabica*	*Babul*
2.	*Acacia auriculiformis*	*Australian Wattle*
3.	*Acacia catechu*	*Khair*
4.	*Acacia senegal*	*Kumta*
5.	*Acacia tortillis*	
6.	*Casuarina equisetifolia*	*Saru*
7.	*Terminalia arjune*	
8.	*Terminalia tomentosa*	*Asan*
9.	*Leucaena leucocephala*	*Subabul*
10.	*Prosopis juliflora*	

India has paid attention to the bioenergy promotion programme in a big way. The concerted effort of the Department of Non-Conventional Energy Sources and Bio Energy Society of India has helped in the identification of problems, causes, symptoms and matching solutions.

Need for greening the arid zones has been felt for long. It can be effectively done by introducing Bioenergy Programmes. However, a lot of interdisciplinary research and development and demonstration work needs to be done.

10

BIOENERGY RESOURCES FOR ISLANDS

Islands under India's territory are almost of the size of Small Island Developing States. All the Island States are heavily dependent on energy imports and similarly the Indian islands are no exception. Increased efficiency through appropriate technology and national energy policy and management measures are commended to benefit the small islands both financially and environmentally.

Andaman and Nicobar Islands are a group of more than 3,000 islands, the great majority of which are uninhabited being too small and with little or no water. The total area of the two groups of islands is 8,349 sq km of which Andamans, with 6,340 sq km, accounts for more than 76 per cent of the land area; and Nicobars accounts for 1953 sq km. The total population of Andaman and Nicobar is 277,989 (updated as per the 1991 Census). The Islands have enough landmass to undertake energy plantation.

Lakshadweep Islands are a group of 36 islands having a total landmass of 32 sq km. This, smallest Union Territory of India, is an archipelago consisting of 12 atolls, 3 reefs and submerged banks. Only 10 of the 36 islands are inhabited with a total population of 51,681 (updated as per the 1991 Census). Thus there is hardly any landmass to grow terrestrial biomass. However, if one were to consider the lagoon area of 4,200 sq km, Lakshadweep has 20,000 sq km of territorial waters. Aquatic biomass provides immense potential for the development of Lakshadweep.

Forest cover in the Andaman and Nicobar Islands is 7,130 sq km. This has prompted wood-based industries to come up. Padauk and Gurjan

---

Fifth Annual Convention, BESI, Vadodara, 30-31 October, 1988.

are among the most valuable timber. There are 857 small scale and handicraft industrial units besides a few large units. The products comprise sawn timber, commercial plywood, match splints and veneers.

An island is a small landmass engulfed by marine environment. India has two major groups of islands — Andaman and Nicobar and Lakshadweep. The sub-soil water is saline which makes conditions totally different from the mainland. This changes the pattern for vegetative growth and productivity. The cost effective large energy infrastructures ideally suited for mainland applications normally are unsuitable for the islands. Moreover, availability of raw materials and resources is also a difficult proposition as the only means of transportation is over the high seas.

Marked Similarity in Energy Consumption Pattern with Rural Areas

Whereas fossil energy like petroleum, coal and natural gas is transported from the mainland to islands appropriate and efficient utilization of locally available resources is equally important and environmentally benign. In this respect solar energy and biomass energy are two important sources locally available on small landmass of India's islands. The energy production and energy utilization in islands will have marked similarities with the energy consumption and use in rural India and coastal India.

The government has emphasized on providing massive inputs to meet a small fraction of the total energy needs of the rural masses as these are supposed to bring in large social benefits in terms of opportunities for community development. This refers to common facilities like streetlights, drinking water, etc. However, little effort is made on improving energy availability in sectors like agriculture, transport, etc. But, there is a focus on utilizing locally available and environmentally benign sources of energy.

Biomass Use

The domestic energy needs, if taken at an average of the Indian rural sector, amount to roughly 25,000 tonnes of firewood for Lakshadweep, and 139,000 tonnes of firewood for Andaman and Nicobar. If emerging energy plantation techniques and genetic planting material is used, then the fuelwood needs can be fully met from the landmass available in the Andaman and Nicobar Islands.

To work out the exact cooking energy requirement, it is necessary to study the local energy use habits in relation to traditional requirements.

Many of the people have other cooking gadgets and the cooking habits are also quite different. Such studies shall have to be close to the ground realities and should not emerge out of the official data. Care should also be taken not to reflect that the energy consumption in the Islands is the lowest. As a matter of fact, electricity consumption in the Islands is 63 kwh per year capita which is more than the electricity consumption figure of Assam, Meghalaya, Manipur, Tripura, Nagaland, Mizoram and Sikkim which are all parts of the mainland.

It is not sufficient to focus only on the domestic sector. Even the income generating sector can have its energy needs met through biomass use. Study of energy use habits must include cooking energy needs and efficient improved techniques; agricultural needs and possible effects of biological fixation of nitrogen and spread of monsoon; and efficient use of energy in agriculture through mechanization. In addition, energy needs in the fishery industry and the transportation are equally important.

At the same time conversion technologies now emerging on the mainland should be effectively transferred to the islands for their industrial, income generating and domestic energy requirements. Such end-use devices as gasifires, starling engines and improved wood-burning stoves and ovens prove extremely advantageous. Building these devices out of locally available material also requires technical and financial support for adoption of already developed end-use devices. This effort will also require about 10 to 12 years of continuous and infrastructural support backed up by financial incentives to achieve optimum results.

Biomass has a tremendous potential of meeting energy needs in the Indian islands of Andaman and Nicobar and Lakshadweep.

11

EMISSION MONITORING OF WOOD BURNING STOVES

Studies on the energy consumption pattern in rural India have established that about 85 per cent of the domestic energy is consumed only for cooking. This comes primarily from biomass — woody or green. Fuels include fuelwood, animal dung, crop residues, etc., and are normally gathered cost-free by the people who consume them and, therefore, are categorized as the non-commercial sources of energy. Since this all is being done on a non-commercial basis, nobody has attempted to determine the efficiency of use or the associated emission hazards. However, at a juncture when the Bio-Energy Society of India is holding its First Annual Convention and Symposium and trying to focus attention on energy from wastes and wastelands, it would be appropriate to discuss the efficiency of use and the emission pollutant control in domestic applications.

Fulfilment of energy needs of the rural area is presently affecting our eco-balance adversely. Whereas in major parts of the country, villagers have the tendency to pluck farm residues of cereal crops like pulses and use them as firewood or buy wood from the open market, the situation in the remote and hilly areas is different as they normally have horticultural crops and people have to spend 4-5 hours a day in the collection of firewood.

The Department of Non-Conventional Energy Sources has launched a major national project on demonstration of improved *chulhas* for helping to meet the cooking energy needs. The project is designed as a programme for women to mitigate their drudgery and reduce their health hazards and also save fuel in the process. To promote these activities, government support is furnished and the targets laid down emphasize that

---

First Convention of BESI, New Delhi, October 14-16, 1984.

5 lakh improved chulhas would be installed by the end of 1984-85 and 5,000 villages would be made smokeless. The programme is expected to pick up further as more models of the improved chulha become available. One of the public sector undertakings has been assigned a target of one lakh improved chulhas for installation in the houses being constructed by them for the economically weaker sections.

When we turn to other parts of the globe, it is observed that the recent growth in the use of biomass as fuel in the Western world has resulted in concern over the environmental effect. In those Western countries where space heating is important, use of wood is becoming popular. A little known fact is that even in the USA since the first oil crisis, use of wood as a fuel has grown much faster than that of any other energy source — 7 per cent per year since 1973.

Recent reintroduction of biomass as a source of energy in the developed countries has created awareness of the likely environmental pollution. Certain standard products for air pollution monitoring are being extended to this field. The basic assumption is that under small-scale combustion conditions, biomass fuel has a significant emission factor for several important pollutants.

Whereas biomass is used for space heating in such a fashion that there is no draft affecting the inhabitants, its use in small cookstoves or *chulhas* in the developing countries normally leads to adverse effects on the health of the residents of the area. It is with this particular concern that efforts are now being made to study the emission characteristics of biomass fuel heating stoves of the type commonly in use in the developed countries. This is becoming more of a concern as outdoor smoke levels rise in communities relying on such appliances. From the wood smoke studies, a number of organic compounds have been characterized in the hydrocarbon portion of the emission. Indeed, several hundreds have been identified many of which are polyaromatic hydrocarbons that have been shown to be mutagenic or carcinogenic. Although the problem of smoky village kitchens has long been noted by observers in the rural areas of the developing countries, they are still not aware of its adverse effects on health. The human exposure to several important pollutants that can be considered as a result of these concentrations are of the magnitude higher than the typical urban exposures. As a matter of fact, the procedure adopted for studying urban pollution may not be totally applicable to biomass smoke study. Urban particulates, for example, are usually associated with sulphur oxides because of the composition of their principal source — fossil fuels. In areas polluted by biomass smoke, however, particulates are usually associated with carbon monoxide and hydrocarbon vapours and droplets. There are further important differences between typical fossil fuel smoke and biomass smoke in the size distribution and economic nature of the aerosols, percentage of element carbon, content of trace metals, etc.

Four Pollutants in Biomass Smoke

Dr. K. R. Smith of the East West Centre, Hawaii, has carried out extensive tests on tobacco smoke and is attempting to set analogous standards. According to his observations and data, four pollutants are found in significant amounts in the biomass smoke of all kinds: respirable particulates, carbon monoxide, formaldehyde, and particulate benzopyrene. Each of these has been the subject of considerable attention in its own right as a health damaging pollutant. Each of them also represents an important one of the four principal classes of pollutants found in biomass smoke particulates, gaseous hydrocarbons and PAH. To test this analogy in a quantitative manner, cigarette and wood fuel would have to be studied for relative emission factors, area concentrations and normal human doses of these four pollutants.

To understand the relative health effects of pollutants, it may be appropriate to measure the dose observed or deposited in the body. However, there is variability in the many air contaminants deposited or observed by different people at different times. The breathing rate where mouth or nose breath is occurring, and the condition of the respiratory system all affect deposition. For example, in cigarette smokers there are the additional variables of smoking behaviour. If the smoker inhales the smoke and smokes the butt down to almost nothing, the dose per cigarette is going to be much larger than that of a normal smoker.

Although wood smoke and tobacco smoke have many similarities, there are also differences. The temperature of the cigarette smoke, for example, would normally be higher. There may be some sort of saturation effect at the high concentrations, experienced by the active smoker, leading to lower deposition rates per g of material inhaled; on the other hand, the hot dense smoke from smoking may inhibit or damage natural lung clearance and other defence mechanisms to the extent that deposition efficiency is higher with such exposures. However, it may be that 95 per cent of breaths that are low exposures for the active smoker allow the lung defence mechanisms to operate more efficiently than they can when every breath contains significant concentrations, as in the case of *chulha* smoke.

Dr. K. R. Smith has also observed that of the four pollutants examined, it seems that a nominal dose of two of them is roughly similar for cigarette smokers and village cooks — formaldehyde and carbon monoxide. In all cases, village cooks receive higher nominal doses than passive smokers. On the basis of these comparisons, it may be expected that the health impacts among the village cooks would lie somewhere below those for active smokers and well above those for passive smokers. It should be mentioned, however, that many other pollutants are not included. Nicotine, in particular, would seem to be nearly absent in wood smoke and yet it is an important health damaging pollutant in tobacco smoke. Tests under specific operating conditions in rural areas will have to be conducted to achieve satisfactory designs. Standards for the same are being worked out to lay down procedures for testing in laboratories such parameters as efficiency, gas emissions and particulate size. However, for day-to-day production, all these tests are not necessary and only some of them may be undertaken. These tests help in arriving at the right type of designs for different fuels and sites.

Further, Dr. Smith has tried to develop a parallel from cigarette smoking. This does offer a clue to the potential danger but does not give exact data on the same. It is necessary, therefore, for the Indian experts to conduct their own tests and reflect their findings appropriately.

12

FUTURE GLOBAL PERSPECTIVE IN COOKSTOVE DESIGN

Woodstove use in India contributes to providing all the necessary end-use system for domestic cooking in most of the rural, coastal and urban poor segments of the society. Designs of cookstoves have varied from highly inefficient almost open - fire devices to highly efficient totally enclosed woodstoves. Improvement in efficiency has been recorded in different laboratories but extension of highly efficient woodstoves has not been an easy assignment. Very often the users in the field come out with data and figures contradicting claims of the scientists.

Factors Affecting Efficient Cookstove Use

Fuelwood remains the main fuel for rural communities in many developing countries; and for urban communities too, where people do not have access to, or cannot afford, more desirable alternative cooking fuels. In remote rural areas, especially in poorer countries, modern fuels are virtually unobtainable in any substantial quantity. With the communities dependent on traditional fuels, fuelwood is the preferred fuel, but in regions where wood supply is scarce twigs and leaves may be used as fuel and in some countries other biomass such as crop waste and cowdung may be used as fuel. This is particularly the case in the Indian subcontinent where 50 per cent of biomass energy is from crop waste and dung and in certain areas of Africa. Various factors having a bearing on efficient cookstove use are

---

International Training Course on Technology, Management and Development of Improved Cookstoves, organized jointly by FAO, MNES and IIT, Delhi, New Delhi, March 25, 1993.

given below:

(i) In the changing global development scenario rapid urbanization is taking place in rapidly developing countries. Such a shift brings more people into the urban centres and with them come the cooking habits. This migratory population, creating urban slums, continues cooking on fuelwood. On the other hand, many of the developing countries now opening out to market economies and commercial agriculture may seek higher export of agricultural produce on the one hand and more production of high value, agricultural products for urban centres on the other. Such a situation has often resulted in small and marginal land owners these becoming farm labourers and majority of the people shifting from superior grains to coarse grains. In both these cases there will be a corresponding change in the cooking habits;

(ii) Over the next 30 years, rural population dependent on fuelwood is likely to increase by more than 100 per cent. Corresponding increase and surplus regenerative fuelwood quantities are not likely to be available, a major reason being space required for human settlement and for producing food. This situation does not mean that the quantity required for cooking in terms of useful energy will reduce drastically in the rural areas. It was, however, required that the efficiency of use gears up proportionately so that the fuelwood use elasticities may come into play and the same quantity of useful cooking energy is available from a lower physical quantity of fuelwood;

As an example, the average domestic cooking energy need in the rural areas in India has been observed to be 425 k calories per capita per day, being obtained from 1 to 1.5 kgm of woody biomass, depending on whether the fuel was bone-dried fuelwood or agro-residue. The present average consumption of cooking energy is so low that any reduction in useful energy is ruled out, but the physical quantity of consumption can be reduced by improving the use of efficiency;

(iii) The programme of improved cookstove, using low cost, local materials is being carried out in many developing countries with the objective of reducing consumption of fuel for cooking and heating and at the same time reducing smoke in the kitchen and improving kitchen hygiene. Improvement in efficiency is important for women. With the decreasing forest land, women in rural areas will have to face bitter hardship in firewood collection and will, therefore, try to impose their authority regarding selection of cookstove technology to reduce the physical drudgery involved in collection of fuelwood;

(iv) As a result of higher agro-productivity levels in more countries, corresponding agro-residue availability will also increase requiring designs suited for use on a wide variety of fuels rather than on specific fuelwood alone;

(v) Intensive industrial activity in the building sector and the trading will also call for higher use of timber. That will again leave large quantities of residue from the sawmills. Generally, such wastes are used for power generation in low efficiency small generators. But, in certain areas, this residue may also find use in the domestic sector in the form of briquettes;

(vi) Similarly, briquettes are also gaining popularity. A number of industries have used a wide variety and range of agro-residues for densification and thereby production of briquettes that have greater energy concentration per unit of weight and burn clean. Under the administered prices of fuel in India briquettes are observed to cost less than even coal in some cases;

(vii) Charcoal making is quite prevalent in certain countries. Thailand and Senegal have a strong tradition in the use of charcoal as domestic fuel while in others it plays a very little small role. Because of its greater convenience and greater energy concentration per unit of weight, charcoal is sometimes favoured for transportation to urban markets. It is also favoured as it is cleaner and a more efficient fuel;

(viii) Environmental considerations are yet another reason for improvement in technology mainly from the use point for reducing the emissions rather than for improving efficiency;

(ix) It may thus be observed that there will be a need for improvement in designs based on the perceived needs of the users, on the one hand, and of scientists and experts on the other. The non-government organizations and women's groups involved in cookstove development and dissemination, for over a decade now, have gained an insight into the problems and are quite geared for the innovative changes;

(x) As the incidence of poverty and illiteracy reduces in the developing countries, information dissemination will play an important role in accepting newer versions. Acceptability of modifications and designs by the public will become easier, but a lot more professionalism will be required in the extension agencies as the users would be ready with many more queries about modifications;

(xi) The number of persons in the rural areas will increase substantially, but there will not be a proportionate increase in the number of cookstoves in view of the wide-scale dissemination of biogas technology and its acceptability on improving reliability in its performance; and

(xii) Charcoal and briquettes will make a deeper market penetration in view of the commercial organizations having entered the field of production, and large rural migration to the urban centres will necessitate their access to certain fuel which looks somewhat similar to the fuel used by the urban users. Such populations will certainly avoid using fuelwood as that does not give them an elitist look.

The above mentioned factors need to be analysed properly to develop improved cookstove design concepts. Some of the issues to be considered include the following:

— Concentrated charcoal or briquettes will be preferred.
— Low polluting designs for urban areas will be preferred.
— Rural population to travel long distances and therefore desirous of consuming low quantities.
— Improvement in efficiency will be emphasized.

Getting the same quantity of useful energy from a lower physical quantity will involve following issues:

— Larger Agro Production — More Agro-Residue.
— Designs to suit them.
— Government's shift in priority and lower allocations for core support. Also resulting in lesser control and more freedom to develop designs.
— NGOs and private funding agency involvement to promote large number of designs.
— Fulfilling needs of disabled and special groups.
— Reliable designs based on market demand.

Emerging Improved Cookstove Design Concepts

Collaborative research programmes relating to the changing agricultural pattern and eating habits will have to be undertaken. There will be a basic shift from major R&D institutions and educational institutions designing and developing new models to non-government organizations developing new designs at the operative level and disseminating them. The role of R&D institutions will now be restricted to study major aspects in reducing emissions and thereby preserving environment, and on developing new materials for cookstove manufacture.

Future design improvement will be "user friendly" as these will emerge at the grassroot level. Also, the initial phase of cookstove designs has created an awareness in the rural areas that will be helpful in the faster development of a design using local materials and traditional skills.

The emerging global scenario in the cookstove design can be summarized as follows:

Target Groups:

— Rural
— Migratory rural-to-urban population

Operating Medium:

— Firewood
— Agro-waste
— Charcoal and Briquettes

Models:

— Portable models for commercial production without subsidy on the sale of models, but with support to manufacturers for standardization, testing and marketing.
Models to be Developed Commercially:
— Fixed models on site specific needs by NGOs, women's organizations and local masses by using and improving local materials and local skills.
Programmes with support from Private Funding.
— Designs to suit environmental requirement in urban setting.
— Fuel Saving — The major concern in rural areas.

Supporting Development:

— Standardization and production of cooking vesels giving energy efficiency in use.
— R&D activities involving extension agents.
— Development of cookstoves, biogas burners, etc., to be interlinked, for efficient use.

It is not suggested that all future development will consider only the above-mentioned factors. But, some of the above-mentioned factors will certainly guide the development programme at different sites.

13

IDENTIFICATION OF BIOENERGY INFORMATION CENTRES

Recent advancements in the energy sector have offered a whole range of energy mixes, optimally suitable for different geographic sites. The developing countries, by and large, are ideally suited for the exploitation of renewable sources of energy. India is no exception and due to its ideal location, it has the advantage of being endowed with different sources of energy, like solar, biogas, biomass, wind, ocean, etc. These offer a wide variety of possibilities starting with their applications directly suited for industrial needs to small rural energy needs. The most important aspect is the possibility of using these sources directly in the rural areas.

The field is, more or less, new and some of the technologies are fully matured and others are in the process of maturity. The nature of technological work is as diverse as the possibility of arriving at their maturity. Wood-burning stove is an example of how designs can be adopted by the rural people in the matter of a month or two. On the other hand under the energy plantation programme, where biomass is grown for ultimate gasification to ascertain the possibility of its wider application, gestation period is long. Even if we assume that the land is available and all the infrastructure is groomed to a stage where we can easily take off within a month of our decision-making, the first crop will be available only after four years and thereafter every year with a cropping cycle of three years. This is considered an ideal solution. Thus, it will be noticed that a minimum of five years of gestation period is necessary before cost

---

Seminar on Bio Energy Information Systems, organised by the Society for Information Scientists, New Delhi, August 27-28, 1985.

effectiveness of a bioenergy resource can be practically ascertained. Moreover, this programme has so many social implications that no ideal model can be prescribed and then replicated in different parts of the country. Thus, energy application in geographically different situations will be a new experiment which will, in principle, broadly refer to a national plan, but will tactically be different in every situation.

Renewable Energy Utilization

In seeking to increase the use of renewable energy, India has to pay full attention to three factors: First, rapidly evolving technology since renewable energy technology was neglected in favour of fossil fuel technologies during the years that cheap petroleum was available. Second, judicious selection has to be made amongst the broad range of applications, from technologies that vary in sophistication from photovoltaic cells to clay stoves or chulhas. Third, many of the rapidly changing technologies that they must become acquainted with and match to their needs were developed to meeting needs other than their own.

To speed up this advance, the government has encouraged research, development, demonstration, and commercialization of renewable energy technologies and applications. They range from the applications of well-known principles of science and engineering to explorations at the frontiers of knowledge. Since growth has to be rapid, selective import of technologies and products is also considered. However, the experience has shown that technology imported from the industrialized countries seldom matches our needs; it requires major changes in scale and extensive adaptation to suit local conditions.

We should, therefore, realize the amount of work required and the sharpness necessary to differentiate between the needs of different sources and devices and the degree of maturity of the technologies that have been developed.

Biomass Availability and Utility in India

Biomass is a potential source of energy where attention is being focussed. The total amount of biomass produced, per annum, in India is about 1,249 mt; (energy equivalent of biomass as 4,400 kilo calories per kgm) the total latent energy comes to about 5496×10^9 kilo calories. This latent energy availability can be increased by improved photosynthesis and also by developing species which can thrive on saline soils and under unfavourable environment.

About 40 per cent of this potential can be used as a bioenergy resource. This will be equivalent to 350 mt of firewood or 175 mt of kerosene.

Contribution of biomass can be significant in time to come. Research work done recently has helped in identifying the species of quick-growing trees and crops.

The alcohol produced in the country is hardly sufficient to meet the demand of the chemical-based industries. In view of this situation, the government has decided to explore alternative sources of liquid fuel. Results of experiments on the enzymatic production of ethanol from cellulosic hydrolysates are encouraging.

Gasification of biomass is the other route. Gasifiers of 5 kw to 40 kw have been in operation. Energy plantations have been planned to produce 3 mw from 1,000 hectares of plantation.

Talking of biomass as such which is the basic concern of today's workshop the present-day research, development and extension has four distinct areas that need to be properly considered. These are as follows:

(i) Assessment;
(ii) Production;
(iii) Processing; and
(iv) Utilization.

Categories of Information Seekers

All these require proper information availability, processing and application to arrive at the right inference, so that right decision-making is possible. The information has to be processed for three different groups as follows:

(i) Information for use of scientists, industrialists and academicians working in the field of bioenergy;
(ii) Information necessary for extension agencies, that is, those who are involved in the promotion of its wider application amongst the end-users; and
(iii) Information for those who are involved in writing about bioenergy and creating an atmosphere conducive for development and growth of this sector. This also includes the mass media.

While deciding about these broad areas the first one has a very basic requirement and will also have to deal with various aspects of the national energy scene — different energy sources available and the gaps that need to be filled in. At the same time it will also have to identify whether bioenergy resource can either substitute or supplement the fossil or other existing sources of energy. As far as the commercial supply sector is concerned there is a certain system for systematic collecting, processing and disseminating information relevant to decision-making needs. These sectors have developed a fairly comprehensive database concerning their energy resources/results, production technology use and sales/distribution of their products to different sales outlets/buyers. The data reported in their conventional reporting units as the products of energy sector are heterogeneous from the physical viewpoint, for example, electricity, solid, liquid and gaseous fuels. A recent report of the Advisory Board on Energy observes that "the current energy information concerning the end users, process sale, efficiencies, supply continuity, supply availability (quality, quantity and dependability) is far from satisfactory. Most of these data are not covered as these are not needed by the supply sectors for their internal operational control, nor does there exist a centralized agency on total energy management which would prescribe such information. Hence the data gaps arise due to the lack of structured energy information system for the total energy sector.

Another difficulty concerns the gaps in the desired information. Many public statistics provide data which could not be obtained in time and which, therefore, had to be estimated. However, very few statistics provide the base for such estimation or do not replace estimates with the actual data when made available.

The status of data for the non-commercial sector is even worse. There is virtually no quantitative evidence on the consumption of non-commercial fuels regularly collected and updated by centralized agencies, although these fuels account for half of the total energy consumption in India. Data on consumption and usage losses and efficiencies are available only through surveys which are neither regular nor cover all geographical areas and data elements.

Different Sectors for Information Gathering

The information on bioenergy has to be collected in four different sectors. These may be discussed as follows:

Some of the information that is extremely necessary to start with on the assessment of resources includes the technique for an overall assessment of the biomass cover in the country. This is difficult to carry out by ground level surveys in a short time. The easiest course is to follow the remote-sensing technique to get a quick survey report and have the first assessment on this basis. This data then further needs to be cross-checked and assessed on the basis of physical verification and ascertaining of the suitability of land for energy use. There can always be a difference between the biomass cover and the biomass useful for energy exploitation.

Projected additional cover to be provided by on-going programmes and schemes of various Central and State Governments needs to be taken into account to forecast biomass availability over short-term and long-term periods.

In addition, a proper assessment of agricultural production and the availability of wastes like bagasse, rice husk, etc., will also make it possible to assess the quantity of total generated biomass and the available biomass in the form of wastes and residues for our use. The quality of biomass then will have to be assessed and grouped according to the calorific value of different species.

The second stage is the production of biomass. Information on this particular aspect will have to be linked directly with the agricultural practices and the forestry programmes being followed in the country. Forestry, as we are all aware, is being adopted under the "Afforestation" programme as well as the "Social Forestry" schemes where the rural masses are directly involved in growing small quantities of firewood in or near their own lands and households. The Energy Plantation for Power Programme of the Department of Non-Conventional Energy Sources is the third category which will lead to biomass production and must be taken into account.

Thus, the possibility of production will, on the one hand, depend on the existing knowledge of agriculture and forest produce, and on the other on the emerging processes for better germination, higher productivity per unit land and better calorific value produce. New germ plasms, tissue culture and other bioengineering practices that are emerging and are expected to revolutionize the whole plant growth pattern will have to be properly evaluated and information made available on them. This will, as a matter of fact, be carrying us directly to the frontier areas of science and technology.

The next stage is of information availability on processing techniques. Different types of biomass available to us can be converted into solid, liquid and gaseous states by briquetting, liquification or gasification techniques. All these, though known for a long time, were never evaluated in terms of their flexibility of application ranging from a few kilowatts to megawatt stage and using different varieties of biomass of varying heat values. Though these technologies already developed and used on a limited scale were known to us for a long time, they have now to undergo a revolutionary change to directly meet the needs in the present situation. Thus the information varying from the existing to capability under different situations needs to be properly understood.

The last of the areas where information is crucial is the information on utilization of bioenergy resource. In view of the versatility of conversion it can be suitably adopted for industrial, transport and domestic applications. The end resources thus are not a closed group but they vary from the common man in the street to a highly sophisticated industrial set-up in any of the technologically sound business group. Therefore, the information on the utilization part will require a wide range of social, economic and geographical considerations to be encompassed in the system.

Bio-Energy Information System

The Department of Non-Conventional Energy Sources may consider setting up a "Bio Energy Information System" with a Focal Institute and a number of decentralized Centres. This system will provide descriptive information and numerical data on bioenergy. The system needs to be geared to the needs of policy planners, industrialists, decision-makers, research workers, administrators and the public. The distributed network of database will lead to ensuring integration of national efforts in bioenergy information collection, storage, retrieval and wide dissemination by avoiding duplication of efforts.

This network has to be independently set up directly under DNES. Linkages can be made with facilities like INSDOC, and ENVIS by setting up collaborative projects at these Institutions or Centres. The financial support will link these major institutions to the Bio Energy Information System.

While deciding on centres DNES may have to take major "first ever" steps for which precedents do not exist.

These Centres must be linked to R&D institutions, academic institutions, industrial houses and voluntary extension agencies. DNES will have to consider promoting these Centres under industrial houses or connected with major bodies like FICCI, ASSOCHEM, AIMO or AIEI (now called CII). Such Centres are to identify the needs and status of industrial technology.

Thus far we have discussed only the information for scientific purposes. The programme has also to serve policy framers like parliamentarians, government officials and mass media personnel who are non-technical but have a say in decision-making. Regular information bulletins must be issued on the advancements and needs, with economic and special benefits. On the other hand instant retrieval and reprographic services should be available at the institute.

14

BIOENERGY UTILIZATION — Political, Social And Technological Issues

Major challenges facing the world community today are the food demand of an increasing population and reversal of environmental degradation. Each depends on the other. Without the healthy, productive environment, it will not be possible to feed the world or provide it with the energy they need for their very survival. The world population which stands at about 5 billion is likely to rise up to 6 billion in the year 2000, and 8.5 billion by the year 2025. Most of this growth will take place in the developing countries where satisfying the increased demand for food will require an estimated 60 per cent expansion in agriculture. Already some 500 million people suffer from malnutrition and 15 million die each year from starvation and related diseases.

The solution has to be searched in a global context and examples are to be picked up from similar countries. Fuelwood is still an important source of energy and the foreign exchange crunch requires development of bioenergy resources for solid, liquid and gaseous fuels.

Fuelwood for Energy Needs

The worldwide rate of deforestation is now estimated at more than 17 million ha a year. Clearing land for agriculture accounts for most of this; next comes the demand for firewood. No doubt, forest conservation is an investment in the future. Forests are the reservoirs of biological diversity which holds the promise of improving crop production as well as

Seventh Convention of BESI, New Delhi, July 12-13, 1991.

providing methods for adapting to global warming. Forests themselves may limit or soften the impact of climate change.

Trees have played an important role in the wellbeing of the poor. Trees have been traditionally a source of food, fodder and fuel. Sustainability depends mainly on relevant management and conservation of the natural resource base and the orientation of technological and institutional change in such a manner as to ensure the attainment and continued satisfaction of human needs for present and future generations.

Fuelwood is the main source of energy in the developing world; more than two billion people are dependent on wood and charcoal for cooking and preserving their food. Fuelwood provides 93 per cent of all energy in Burundi, 84 per cent in Haiti, and 97 per cent in Bhutan. In the developing countries as a whole, eight times more wood is used for fuel than is logged for industrial purposes. In the year 1980 nearly 1.3 billion people could not meet their fuelwood needs. By the year 2000, nearly 3 billion could face shortages. In India also, fuelwood meets the basic domestic cooking needs of 85 per cent of its population living in rural areas.

Very little do people realize that the human health often depends on access to fuelwood. Cooking releases the nutrients in some foods while others can be poisonous if not prepared properly. As the wood becomes scarce, people eat lesser quantities of cooked food and the food is cooked for shorter times. They also have less fuel for boiling water, essential for disease prevention.

Managing existing fuelwood supplies is an important first step in meeting the present and future needs. Biomass resources can also be used not only for the woody biomass but for liquid and gaseous fuels also. This role is not going to diminish in the nearby future or even in the long-term as the economic situation is globally driving the developing countries to switch over to fossil fuels.

Neglect of Bioenergy in Energy Policy

Energy policy and plans in the developing countries should mainly focus on the domestic rural requirements. But, that is not true. As the countries gained independence they carried bureaucratic and administrative legacy of their rulers. The agricultural sector very often remained neglected as the energy availability in terms of commercial sources is negligible. The other problem is the inherent difficulty of data collection and management; dispersed and non-monetized nature of energy consumption patterns in the rural areas also contribute to this neglect. Energy authorities rarely have institutional or operational mechanism in the rural areas. On the other hand, most rural and agricultural development plans and programmes do not explicitly treat rural energy requirement, because of lack of technical capability and awareness of options.

Understanding Nature of Biomass Fuels

Biomass fuels can contribute signficantly in meeting rural and urban energy needs. However, to effectively fulfil this role, the donors, development planners, decision-makers, environmentalists and research scientists must afford significantly more attention to understand the comprehensive nature of biomass in climate regulation, food production, water resource availability and soil fertility besides energy provision.

A large number of examples can be picked up from different parts of the globe especially from the Amazon region and the sub-Saharan Africa to reflect what can happen if the biomass resources are mismanaged. Such an act had already resulted in ecological degradation and human suffering in a very short time. Mismanagement includes harvesting timber without replanting, conversion and unsustainable use of land for crops or livestock production and other imprudent practices.

At the same time introduction of new sources of energy or new techniques in technologies without going into their financial considerations can be fatal.

Major Case Studies of Biomass Fuel Utilization

Brazil's "Energy Substitution Programme" based on large- scale ethanol production and consumption has long been acclaimed by many as an outstanding success. Besides pointing to its present short-term supply problems there are structural and long-term distortions which may be introduced into the economy and society. The alcohol programme in Brazil was started in 1975. As a first target, the programme was designed to guarantee use of a mixture of anhydrous ethanol and gasoline, up to 22 per cent of the former by volume. By the year 1979, a second target started by the production of straight ethanol fuelled cars. Since then, the programme has been able to achieve all official goals set for volume of ethanol necessary to attain market requirements. As per the 1989 demand, approximately 4.2 million straight ethanol cars and 6.5 million of 22 per cent blended ethonol/gasoline cars were used. The extremely low oil prices in the last few years caused the ethanol price to be valued below its national average production cost with the consequence of stagnation in the amount of land use for the sugarcane crop. Simultaneously the favourable

1989 sugar price brought an extra difficulty to the ethanol market since it was still possible for half of the distilleries to produce sugar or ethanol. Market pressure tempted a few of them to direct their sugarcane to sugar production. Stagnation in sugarcane production and better sugar prices coupled with the continuous increase in the demand for spirit ethanol resulted in ethanol shortage. The solution to this problem was urgent to guarantee 4.2 million car owners' right to assured supply of ethanol and the credibility of the programme in future.

The survival and expansion in a logical way depended on diffusing the ethanol cost and simultaneously opening more space for it in the Brazilian energy profile so that another energy form from the same raw material is possible. There are a number of arguments in favour of production of electricity from sugarcane bagàsse which is an ethanol by-product. Electricity could be so produced by sugar mills and supplied to the electricity supply agencies.

Another example is "Energy Plantation for Power" programme of the Philippines launched with great fanfare. The pilot programme launched on 1,000 ha of land was planned to produce 3 mw of electricity from a centralized system. In the process of plantation, 100 persons were directly employed (whose family energy needs were also met through the plantation). The cost of electricity produced from such a plantation was estimated to be lower than the local power generation cost. Everything went fine with the project so long as the woody biomass around the centralized incineration plant was used. As the distance of biomass source increased, the transportation cost became prohibitive and the cost of power generation became almost double.

These two experiments, held outside India, have been chosen to demonstrate that "technical feasibility" alone is not sufficient to decide on large-scale substitution of energy source. Their implementation has to be preceded by thorough perspective studies of the social and economic impact, as well as of their larger political implications. Ethanol production has also resulted in certain distortions in the Brazilian economy the most important being the regional concentration on ethanol. The ownership of the land has been strongly concentrated with the elimination of numerous small owners and farmers.

Therefore, any decision about the biomass energy programme where biological process and society involvement are at a stake will call for appropriate decisions in financial and administrative terms also.

Biomass Energy and Rural Employment

There is a direct relationship between bioenergy utilization and employment in the rural areas. This has a greater significance in a country like India where technolgoies are now available for a large number of village based-industries including the food processing industries. The rural population, in turn, is also on the lookout for employment opportunities. One-third of the total population falls under the category of youth in India. By the year 2000 when India's population will be one billion, the youth population will be 330 million. About two-thirds of them will be living in villages and looking out for employment and income generation opportunities. Added to this number will be the adult agricultural labour. Under the present economic policies where agricultural production is being promoted for catalysing rural industrialization it is necessary to think in terms of various bioenergy options.

This obviously leads to the conclusion that there are a number of relevant issues which have to be taken into consideration for large-scale dissemination of bioenergy technologies and production activities.

Issues for Bioenergy Promotion

Political Issues

The political issues involve determination and resolve of the national governments to use biomass resources to supplant commercial energy sources in a technologically sound and economically effective fashion. India has a landmass of 329 million ha of which 93 million ha is a total wasteland in the non-forest area and about 43 million ha is the partial wasteland in the forest area. These wastelands are a potential source for growing energy plantation. A decision on the greening of these is a political decision but not independent of technological, administrative and financial considerations. In 1985, the government made announcements for plantation at the rate of 5 million ha per year but later found that was not infrastructurally feasible.

Various elements of the political issues are as follows:

The first element is the reflection of political determination.

The second element in policy has to be whether the government wants plantation to be executed with the involvement of local masses or not — "stop-and-go" arrangement in this regard will not work. Appreciation of people's participation should not simply be a lip- service but an administrative necessity.

The third element relates to long-term landuse plan. Areas will have to be clearly earmarked for agricultural and bioenergy plantation. Allocations will have to be accordingly made for the two sectors. Even within the energy plantation sector the concept of fuel and fodder will have to be promoted.

The fourth element in political decision will involve the government to give proper weightage to the environmental preservation aspect.

Social Issues

Use of landmass and involvement of people in the bio-energy plantation programme is an important component. The society and plantations will be interactive in that 85 per cent of the rural population depends for domestic energy needs on the fuelwood; and in turn the bioenergy plantation depends on the attitude of the people.

The first issue of societal concern is involvement of the people in policy, implementation and maintenance of bioenergy plantations.

The second issue is priorities in landuse, the percentage to be used for agriculture and for bioenergy plantation.

The third issue involves a decision about the preferred route to be followed, i.e, woody, gaseous or liquid.

Technological Issues

Technological issues dominate activities in the bio- energy sector. Plantation is still considered a new activity and only research and development are being promoted. Internationally also only the technological issues are considered. A number of issues relating to the technology come to fore.

The first issue is the creation of an appropriate environment whereby research in identified few areas only is undertaken.

The second issue involves institutional arrangement for availability of inputs like quality seeds, plants, fertilizers and fiscal incentives; and institutional finance.

The third issue relates to training on plantation strategies and processing of biomass.

The fourth issue involves developing "total technology package" on a most cost-effective basis.

These are just a few of the issues involved with political, social and technological aspects of higher utilization of biomass resources.

Macro and Micro-Barriers in Bioenergy Technology Dissemination

The macro-barriers have to be overcome for large-scale utilization of biomass resources. These relate to national level energy supply problems and factors such as policy measures and conventional energy costs. The micro-barriers are normally associated with the daily management, operation and maintenance of on-farm energy plants.

Conclusion

An effective bioenergy utilization programme to contribute to the overall energy scenario in the developing world, especially India, will require deeper involvement of policy-makers in issues relating to political, social and technological aspects. An appropriate approach will call for a holistic approach rather than the present-day technological only.

ANNEXURE

THE EFFICIENCY OF TREES AND PLANTATIONS IN CONVERTING SOLAR ENERGY

The efficiency of photosynthesis sets the ultimate limit on crop productivity (Beadle *et al.* 1985). The plant kingdom is the largest solar energy convertor on this planet. Net primary production each year through the process of photosynthesis is 3 x 10^{21} Joules — about 10 times the world's estimated annual use of all forms of energy (3 x 10^{20}J) (Hall 1984). The amount of energy in total standing biomass (90 per cent of which consists of trees) is about 20 x 10^{21}J or slightly less than the estimated energy in the present fossil fuel reserves (25 x 10^{21}J) (Hall, 1979, 1984).

The solar constant or the average total radiation energy received from the sum per unit of area on a theoretical surface perpendicular to the sun's rays and at the mean distance of the earth from the sun, corrected for atmospheric effects and seasonal changes, is about 440 x 10^{8}kJ/ha/yr. In the tropics and subtropics and at high altitudes more of the sun's incident radiation reaches the earth's surface than at high latitudes, provided cloud cover is low. Photosynthetically active radiation is about 40 per cent of total radiation or about 175 x 10^{8}kJ/ha/yr.

Much of the photosynthetically active radiation is absorbed by plants but only a small quantity is fixed in plant woody tissues which may be subsequently released by burning biomass fuels. The ratio of the amount of photosynthetically active radiation that falls on the plant canopy to the amount of energy fixed by the plants is termed the energy conversion efficiency (ECE). Although the efficiency level is theoretically relatively high at around 30 per cent in terms of gross assimilation (Kok 1973; Kok *et al.* 1976), net assimilation is much less, because energy is lost in processes in the living plant such as respiration. The actual maximum

ECE at the cellular level is probably near 12 per cent. Several comprehensive reviews are available describing the processes and pathways involved (e.g. Jackson and Volk 1970, Hatch 1971, Tolbert 1971, Zelitch 1971, Beadle *et al.* 1985).

Estimates of the ECE of forest systems are few. The maximum solar ECE by woody plants in general is probably between about 6.6 per cent (e.g. Gurumurti and Raturi 1981) and 11 per cent (e.g., Schneider 1973). Hall (1979) has calculated the global ECE in plants systems and concludes that trees make the greatest contribution to solar energy conversion in the world, even though the efficiency of trees is as low as 0.2 to 0.5 per cent on a global basis. Gurumurti (1981) has outlined the causes of low ECE in trees.

An ECE of 0.4 per cent was reported to be very good for temperate tree crops by White and Plaskett (1981), while Szego and Kemp (1973) were of the opinion that an ECE of at least 0.7 per cent could make temperate short-rotation bioenergy plantations feasible and economical.

Frissel *et al.* (1978) described yields of *Populus* sp of 15 dry tonnes/ha/yr which suggests an ECE of about 1.8 per cent. Yields of *Casuarina equisetifolia* obtained by Seshadri *et al.* (1978) had a calculated ECE of over one per cent. Gurumurti *et al.* (1984a) reported a figure of 1.26 per cent for a three-year-old close-spaced Eucalyptus plantation. They also noted two peaks in seasonal solar ECE, one at 18-24 months and another at 30-36 months, reaching 1.77 and 3.26 per cent respectively of photosynthetically active radiation and the peaks corresponding to periods of additional moisture availability for growth (Gurumurti *et al.* 1984a, b).

For three-year-old close-spaced plantations of *Dalbergia sissoo, Cassia siamea, Acacia nilotica, Eucalyptus tereticornis* and *Albvizia lebbek* grown in India, ECE's of 2.09, 1.58, 1.44, 1.26 and 1.19 per cent were calculated based on utilizable biomass yields of 28, 21, 23, 15 and 17 t/ha/yr respectively (Kimothi 1984, Kimothi *et al.* 1984).

Pilot plots of *Eucalyptus camaldulensis* and *E. tereticornis* planted at 0.9 x 0.9 m spacing (over 12,000 plants/ha) and grown for five years in Bangladesh have shown a solar energy conversion efficiency of 5.8 and 4.8 per cent based on above-ground biomass yields of 51 and 42 t/ha/yr respectively (Davidson and Das 1985). The yields reported are unlikely to be as great in routine plantations, because small research plots receive light from the side as well as from above. Realistic expectations would be about half these yields. These citations are included here mainly to show the range of ECE values reported. Without details being made available on the actual site quality and management regimes, ECE's are not directly comparable from species to species and place to place. Only by adopting some average calorific values and the range of productivity of biomass for each of several species is it possible to compare their relative average solar energy conversion efficiencies.

Based on these data and the discussion above, it appears reasonable to state that prospects for commercial success are better if bioenergy plantations are established only with those species on those sites which will yield not less than one per cent and preferably over two per cent solar ECE (corresponding to about 10 and 20 t/ha/yr above-ground biomass production respectively) over rotations of about 5 years. These expectations are tempered by site quality.

Among the better performing species are *Sesbania* spp., some *Eucalyptus* spp, *Casuarina* spp and *Leucaena leucocephala*. The range of values shown can be put into perspective when compared with sugarcane which, ever under very intensive cultivation practices, is observed rarely to exceed 3 per cent solar energy conversion efficiency (Mathur *et al*, 1984)

Environmental and Ecological Considerations in Raising Bioenergy Plantations

As with any crop system, successful tree husbandry depends on several factors among which are:

(i) selecting the most appropriate species (e.g. see Turnbull and Pryor 1978; Webb *et al*, 1989)
(ii) using seeds of good quality (e.g. see Faulkner 1975, Boland *et al* 1980, Zobel and Talbert 1984).
(iii) raising healthy seedlings for transplanting (e.g. see Evans 1982).
(iv) correctly matching species to sites (e.g. see Davidson 1985a).
(v) using proper planting practices (e.g. see Chapman and Allan 1978, Evans 1982).
(vi) promoting adequate aftercare (weeding, protection from browsing) (e.g. see Evans 1982).

Most of these factors have environmental and ecological implications.

Species Selection

Little (1983) described 90 species of fuelwood crops while various National Academy of Sciences (NAS) publications from the USA (e.g. NAS 1979a, b 1980, 1981, 1983a, b, c, 1984) together included over 100 species which have potential as fuelwood/bioenergy crops. Other authors have made partial inventories, e.g. for the Northern Hemisphere (Lavoie and Vallee 1981). In these publications emphasis was placed on species suitable for growing as fuelwood for individual family and village requirements; but species suited to cultivation in bioenergy plantations for domestic fuelwood and charcoal, fuelling small industries and crop drying (e.g. tobacco and tea drying, brick and tile kilns, molasses manufacture) and electricity generation were also included.

Species suitable for bioenergy plantations normally should have the following attributes:

- (i) capacity for rapid growth on a wide range of sites (e.g. pioneer species);
- (ii) ability to produce wood of high specific gravity, even when grown rapidly in short rotations, i.e. high above-ground biomass production per stem and per unit area of land;
- (iii) high wood calorific value (related to ii, above);
- (iv) stem and crown architecture enabling fast growth to be maintained at very close tree spacings over short rotations of about 4-8 years;
- (v) high solar energy conversion efficiency (related to i above); and
- (vi) good coppicing ability.

INDEX